Nonnegative Matrices & Applicable Topics in Linear Algebra

Nonnegative Matrices & Applicable Topics in Linear Algebra

Alexander Graham, MA, MSc, PhD

DOVER PUBLICATIONS, INC.
Mineola, New York

Bibliographical Note

This Dover edition, first published in 2019, is an unabridged republication of the work originally printed by Ellis Horwood Limited, Chichester, England, and Halstead Press: a division of John Wiley & Sons, New York, in 1987.

Library of Congress Cataloging-in-Publication Data

Names: Graham, Alexander, 1936– author.
Title: Nonnegative matrices and applicable topics in linear algebra / Alexander Graham, MA, MSc, PhD.
Description: Dover edition. | Mineola, New York : Dover Publications, Inc., 2019. | Originally published: Chichester, England : Ellis Horwood Limited ; New York : Halstead Press, a division of John Wiley & Sons, 1987. | Includes bibliographical references and index.
Identifiers: LCCN 2019016774| ISBN 9780486838076 | ISBN 0486838072
Subjects: LCSH: Non-negative matrices. | Algebras, Linear. | Matrices.
Classification: LCC QA188 .G712 2019 | DDC 512.9/434—dc23
LC record available at https://lccn.loc.gov/2019016774

Manufactured in the United States by LSC Communications
83807201
www.doverpublications.com

2 4 6 8 10 9 7 5 3 1

2019

Contents

Preface

Although a number of books making reference to the theory of nonnegative matrices exist, most of them treat this topic at a rather sophisticated post-graduate level. The recent rapid expansion in the development and application of various aspects of economic, stochastic and mathematical disciplines which make use of this theory has persuaded me that there is a need for a text which can be understood by readers who are not necessarily expert mathematicians.

To achieve this aim, I soon realised that it would also be necessary to write a background and an introductory survey which included some aspects of matrix theory which may not be well known to the potential reader nor particularly easily found in the required form in the current literature. Chapter 1 is a collection of various topics in linear algebra and closely related matter most of which are used in various forms in the development of the theory of nonnegative matrices. The kernel of this theory is the Perron–Frobenius theorem dealing with the special properties of a matrix having nonnegative entries. Since there is little point in investigating the characteristics of one family of matrices, however simple and beautiful they may be, without an investigation of the properties of some of the other families of matrices, Chapters 2 and 3 are written to do just that. The main theme of the book, aspects of the theory of nonnegative matrices, is discussed in Chapter 4. Chapters 5 and 6 can be regarded as applications of the theory set out in Chapter 4.

Most of the important concepts in the text are illustrated by simple worked examples and Problems at the end of each chapter (except Chapters 1 and 6) with solutions at the end of the book.

Now for a brief outline of the content of the book.

Chapter 1 may be regarded as an introduction to various properties of matrices and determinants. It includes a section on permutation matrices and, in contrast, a section on the theory of graphs, both topics very helpful in the development of Chapter 4. Chapter 2 covers various aspects of the theory of Normal matrices, including unitary and Hermitian matrices. Chapter 3 is devoted to the discussion of some of the properties of positive definite matrices.

Chapter 4 covers the main topic in this book, nonnegative matrices. Section 4.3 deals mainly with Positive matrices and some of the arguments used are repeated, even if in a modified form, when discussing other classes of Irreducible matrices in Sections 4.4 and 4.5. The policy of repeating some of the arguments is intentional and was dictated by my desire to make this topic as comprehensible as possible to the reader. Section 4.6 covers various properties of reducible matrices.

A discussion of M-matrices will be found in Chapter 5. It can be considered as a mathematical application of the results proved in Chapter 4. Finally Chapter 6 is a direct application of these results to Markov chains and stochastic matrices. Obviously stochastic matrices form a subclass of the family of nonnegative matrices and much of the development of this topic is due to our interest in Stochastic matrices.

Worked examples occur throughout the book and serve various purposes; some illustrate important concepts whereas others are applications of these concepts to linear algebra. The applications presented in Chapters 6 and 7 are devoted to various applications of the theory which was developed in Chapters 1 to 5. The final chapter considers a variety of applications and investigations emphasising the great importance of nonnegative matrices and includes some genetic, economic and Markov chain models.

References are given in two forms; [BX] is to book number X in the book reference section and [Y] is the paper number Y in the paper reference section.

Examples of Notation Used

$A(m \times n)$	matrix A of under $m \times n$
$A = [a_{ij}]$	matrix A whose (i, j)th entry is a_{ij}
I	unit matrix
A' or A^t	transpose of A
A^*	conjugate transpose of A
$\mathbf{x}, \mathbf{y} \ldots$	vectors
$A \geqslant 0$	matrix $A = [a_{ij}]$ where $a_{ij} \geqslant 0$ (all i, j)
ϵ	'belongs to'
$\exists$	'there exists'
Z^+	set of all positive integers
Z	either a matrix or the set of integers
$\|A\|$ or det A	determinant of the matrix A
$\mathbf{e}_i$	the unit vector having a unity is the ith entry and zeros elsewhere
$\mathbf{e}$	the vector having all entries unity
$A_{\cdot j}$ or $\mathbf{a}_j$	the jth column of the matrix $A = [a_{ij}]$
$A'_{i\cdot}$	the ith row of the matrix $A = [a_{ij}]$
$A(\alpha)$	the principal submatrix of A obtained by deleting rows and columns (indicated by α) from A
$r(A)$	the rank of A
$q(a)$	nullity of A
$C(\lambda)$	the characteristic function of A, $\|\lambda I - A\|$

tr A	trace of A
$G(\lambda)$ or adj A	adjoint matrix of A
$g(\lambda)$	the characteristic matrix of A, $[\lambda I - A]$
$\lVert\mathbf{x}\rVert$	a norm of a vector $\mathbf{x}$
$\lVert A\rVert$	a norm of a matrix A
$\rho(A)$	the spectral radius of the matrix A
$(\mathbf{x}, \mathbf{y})$	the inner or scalar product of $\mathbf{x}$ and $\mathbf{y}$ $(\mathbf{x}, \mathbf{y}) = \mathbf{x}^*\mathbf{y}$
$U^{(n)}$, $V^{(n)}$	n-dimensional vector spaces

1

Introductory Survey

1.1 INTRODUCTION

This chapter introduces the notation and discusses various results which will be found useful in this book. Rigorous proofs of the various results will be found in books mentioned in the Bibliography, for example [B13], [B14] and [B7]. But non-rigorous proofs and simple illustrative examples are given. The main purpose of this approach is to make the book as self-contained as possible.

1.2 NOTATION

We write the matrix of order $(m \times n)$ whose (i, j)th element (or entry) is a_{ij} as

$$A = [a_{ij}] = \begin{bmatrix} a_{11} & a_{12} & \cdots & a_{1n} \\ a_{21} & a_{22} & \cdots & a_{2n} \\ \vdots & & & \\ a_{m1} & a_{m2} & \cdots & a_{mn} \end{bmatrix} \tag{1.1}$$

If $m = n$, the matrix is *square* of order $(n \times n)$, we will refer to it as the matrix A of *order* n or as $A(n \times n)$.

We write the vector $\mathbf{x}$ of order n with components $x_1, x_2, \ldots, x_n$ as

$$\mathbf{x} = \begin{bmatrix} x_1 \\ x_2 \\ \vdots \\ x_n \end{bmatrix}$$

or in the transposed form as

$$\mathbf{x} = [x_1\ x_2 \ldots x_n]'.$$

The *unit vectors* of order n are defined as

$$\mathbf{e}_1 = \begin{bmatrix} 1 \\ 0 \\ 0 \\ \vdots \\ 0 \end{bmatrix} \quad \mathbf{e}_2 = \begin{bmatrix} 0 \\ 1 \\ 0 \\ \vdots \\ 0 \end{bmatrix} \ldots \mathbf{e}_n = \begin{bmatrix} 0 \\ 0 \\ 0 \\ \vdots \\ 1 \end{bmatrix}.$$

The *determinant* of a square matrix A is denoted as

$$\det A \text{ or as } |A|.$$

We denote the n *columns* of the matrix (1.1) by

$$A_{.1}, A_{.2}, \ldots, A_{.n}$$

so that

$$A_{.j} = \begin{bmatrix} a_{1j} \\ a_{2j} \\ \vdots \\ a_{mj} \end{bmatrix} \qquad (j = 1, 2, \ldots, n).$$

We sometimes also use the notation

$$A = [\mathbf{a}_1\ \mathbf{a}_2\ \ldots\ \mathbf{a}_n]$$

where $\mathbf{a}_j = [a_{1j}, a_{2j}, \ldots, a_{nj}]'$ represents the jth column of A.

Similarly we denote the m rows of A by

$$A'_{1.}, A'_{2.}, \ldots, A'_{m}.$$

Notice that to conform to the previous notation

$$A_{j.} = \begin{bmatrix} a_{j1} \\ a_{j2} \\ \vdots \\ a_{jn} \end{bmatrix} \qquad (j = 1, 2, \ldots, m)$$

is a column vector whose elements form the jth row of the matrix A.

It is the transpose $A'_{j\cdot}$ which is the row vector corresponding to the jth row of A.

The *transpose* of the matrix $A = [a_{ij}]$ is written as $A' = [a_{ji}]$. The complex conjugate of A is written as $\bar{A} = [\bar{a}_{ji}]$.

Rather than write the *complex conjugate of the ordinary transpose of A* as $\bar{A}'$, we use the notation

$$A^*$$

and call it the *conjugate transpose* of A.

For example, if

$$A = \begin{bmatrix} 2 & 1+2i \\ 1-i & 3 \end{bmatrix} \quad \text{then } A^* = \begin{bmatrix} 2 & 1+i \\ 1-2i & 3 \end{bmatrix}.$$

1.3 SUBMATRICES AND MINORS

We consider a matrix $A = [a_{ij}]$ of order n. We denote the set

$$\{1, 2, 3, \ldots, n\} \text{ by } N.$$

Let $\alpha, \beta \subseteq N$ be strictly increasing sequences, known as *index sets*.

We denote the submatrix of A having rows indicated by α and columns indicated by β by

$$A(\alpha, \beta).$$

So $A(\alpha, \beta)$ can be regarded as the 'intersection' of rows and columns of A specified by α and β respectively.

For example, let

$$A = \begin{bmatrix} a_{11} & a_{12} & a_{13} & a_{14} \\ a_{21} & a_{22} & a_{23} & a_{24} \\ a_{31} & a_{32} & a_{33} & a_{34} \\ a_{41} & a_{42} & a_{43} & a_{44} \end{bmatrix} \tag{1.2}$$

if $\alpha = \{1, 3, 4\}$ and $\beta = \{1, 4\}$, then

$$A(\alpha, \beta) = \begin{bmatrix} a_{11} & a_{14} \\ a_{31} & a_{34} \\ a_{41} & a_{44} \end{bmatrix}.$$

Frequently we are interested in square submatrices of A in which case the number of elements in the index sets α and β are equal.

For example, if

$$\alpha = \{1, 2, 4\} \text{ and } \beta = \{1, 2, 3,\}, \text{ then}$$

$$A(\alpha, \beta) = \begin{bmatrix} a_{11} & a_{12} & a_{13} \\ a_{21} & a_{22} & a_{23} \\ a_{41} & a_{42} & a_{43} \end{bmatrix}.$$

A similar definition applies to a vector $\mathbf{x} = [x_1\ x_2 \ldots x_n]'$.

The indices $\{i\}$ of the elements x_i of $\mathbf{x}$ form a subset from the subset N. We write $\mathbf{x}(\alpha)$, $\alpha \subset N$, to denote the vector obtained from $\mathbf{x}$ by omitting the elements of $\mathbf{x}$ whose indices are denoted by the index set α.

For example if

$$\mathbf{x} = [1\ 0\ 0\ {-2}]' \text{ so that } N = \{1, 2, 3, 4\},$$

then

$$x(\alpha) = [1\ {-2}]' \quad \text{if } \alpha = \{1, 4\}$$

and

$$\mathbf{x}(\alpha) = [1\ 0\ 0]' \quad \text{if } \alpha = \{1, 2, 3\}.$$

The determinant of a square submatrix $\det A(\alpha, \beta)$, is called a *minor determinant* or just a *minor* of the matrix.

If A is of order n and the number of elements in both α and β is r, the submatrix $A(\alpha, \beta)$ is obtained by striking out $(n-r)$ rows of A (indicated by α) and $(n-r)$ columns of A (indicated by β). Then $\det A(\alpha, \beta)$ is said to be a *minor of order r of A* or *an $(n-r)$th minor of A*.

In particular, if we strike out the ith row and the jth column (say) of A, the determinant of the remaining submatrix is a *first minor* of A (since $n - r = 1$ in this case).

Similarly in striking out q rows and q columns of A, we obtain $\det A(\alpha, \beta)$, a *qth minor of A*.

If at least one minor of order r is not zero, whereas all minors of order $(r+1)$ of the matrix are zero, we say that the *rank of the matrix is r*. This non-zero minor of order r is called a *critical minor* of the matrix.

If $\alpha = \beta$ so that the indices of the rows and columns struck out from A to form $A(\alpha, \alpha)$ are identical, then $A(\alpha, \alpha)$ is called a *principal submatrix* of A, and

$$\det A(\alpha, \alpha)$$

is called a *principal minor of A*.

We abbreviate $A(\alpha, \alpha)$ to $A(\alpha)$.

For example, one principal minor of the matrix (1.2) is obtained on choosing $\alpha = \{1, 3, 4\}$, it is

$$\det A(\alpha) = \begin{vmatrix} a_{11} & a_{13} & a_{14} \\ a_{31} & a_{33} & a_{34} \\ a_{41} & a_{43} & a_{44} \end{vmatrix}.$$

Notice that the diagonal elements of a principal minor of A are a subset of the diagonal elements of the matrix A.

If $\alpha = \{1, 2, \ldots, r\}$ then $\det A(\alpha)$ (or just $|A(\alpha)|$) is called the *leading principal minor of order r* (or an $(n-r)$th leading principal minor). In this case the diagonal elements of the minor are the first r diagonal elements of A.

For example,

$$\begin{vmatrix} a_{11} & a_{12} \\ a_{21} & a_{22} \end{vmatrix}$$

is the leading principal minor of order 2 of the matrix (1.2).

1.4 EXPRESSING A SINGULAR MATRIX AS A PRODUCT

In this section we discuss one method of expressing a singular matrix as a product of matrices.

In fact, although we shall be concerned with singular matrices, the method is more generally applicable to any matrix A of order $m \times n$, whose rank $\mathrm{r}(A)$ satisfies the inequality

$$\mathrm{r}(A) < \min(m, n).$$

It is assumed that the reader is familiar with a number of important concepts (see [B7] p. 100), very briefly reviewed here.

We associate a matrix $A(m \times n)$ with a linear transformation

$$T : U \to V$$

where U and V are vector spaces of dimension n and m respectively.

(1) The dimension of the kernel of T is called the *nullity of T*, we shall denote it by $\mathrm{q}(A)$.
(2) The dimension of the image space of T is the *rank of T*, we denote it by $\mathrm{r}(A)$.

The *dimension theorem* can be expressed as

$$n = r + q \tag{1.3}$$

where $r = \mathrm{r}(A)$ and $q = \mathrm{q}(A)$. $\mathrm{q}(A)$ is also known as the *degeneracy* of the matrix A.

When considering the product AB of the matrices $A(m \times n)$ and $B(n \times p)$, the rank of AB, $\mathrm{r}(AB)$ satisfies the following inequalities

$$\mathrm{r}(A) + \mathrm{r}(B) - n \leqslant \mathrm{r}(AB) \leqslant \min\{\mathrm{r}(A), \mathrm{r}(B)\}. \tag{1.4}$$

Further, it is assumed that the reader is familiar with elementary operations on matrices and the associated E-matrices (see [B7] p. 238).

Since a typical E-matrix used in this section may involve all the columns (rows) of the matrix it operates on, it is actually a product of what are known as elementary E-matrices. By postmultiplying the matrix by an appropriate E-matrix, we can express any column as a linear combination of all (or some) of its columns.

For example,

$$\begin{bmatrix} a_1 & a_2 & a_3 \\ b_1 & b_2 & b_3 \\ c_1 & c_2 & c_3 \end{bmatrix} \begin{bmatrix} 1 & 0 & -5 \\ 0 & 1 & 0 \\ 0 & 0 & 1 \end{bmatrix} = \begin{bmatrix} a_1 & a_2 & a_3 - 5a_1 \\ b_1 & b_2 & b_3 - 5b_1 \\ c_1 & c_2 & c_3 - 5c_1 \end{bmatrix}$$

corresponds to the operation

$$A_{.3} \rightarrow A_{.3} - 5A_{.1}$$

using the notation defined in Section 1.2. We illustrate the method by considering a number of simple examples.

Let

$$A = \begin{bmatrix} a_1 & a_2 & a_3 \\ b_1 & b_2 & b_3 \\ c_1 & c_2 & c_3 \end{bmatrix} .$$

Assume that $\mathrm{r}(A) = 2$ so that $\mathrm{q}(A) = 1$. It follows that two of the columns, say columns 1 and 2 are linearly independent, and the third column is a linear combination of the first two, say

$$A_{.3} = \mu A_{.1} + \lambda A_{.2} \qquad (\lambda \text{ and } \mu \text{ are scalars})$$

so that

$$A = \begin{bmatrix} a_1 & a_2 & \mu a_1 + \lambda a_2 \\ b_1 & b_2 & \mu b_1 + \lambda b_2 \\ c_1 & c_2 & \mu c_1 + \lambda c_2 \end{bmatrix}$$

We know that we can construct an E-matrix associated with a matrix B such that the product

$$BE$$

has all columns identical to B except for the last column which is a linear combination of all the columns.

Let B be a matrix identical to the matrix A, except for its last column which is zero. By the above argument we can construct E so that

$$BE = A.$$

For the above example

$$\begin{bmatrix} a_1 & a_2 & 0 \\ b_1 & b_2 & 0 \\ c_1 & c_2 & 0 \end{bmatrix} \begin{bmatrix} 1 & 0 & \mu \\ 0 & 1 & \lambda \\ 0 & 0 & 1 \end{bmatrix} = \begin{bmatrix} a_1 & a_2 & \mu a_1 + \lambda a_2 \\ b_1 & b_2 & \mu b_1 + \lambda b_2 \\ c_1 & c_2 & \mu c_1 + \lambda c_2 \end{bmatrix}.$$

Notice that the column of zeros in B does not contribute to the product. If we delete this column in B and the corresponding row in E, so as to make the product conformable, we still have

$$A = \begin{bmatrix} a_1 & a_2 \\ b_1 & b_2 \\ c_1 & c_2 \end{bmatrix} \begin{bmatrix} 1 & 0 & \mu \\ 0 & 1 & \lambda \end{bmatrix}.$$

For example consider the matrix

$$A = \begin{bmatrix} 1 & 0 & 1 \\ -2 & 0 & -6 \\ 0 & 1 & -2 \end{bmatrix}.$$

$r(A) = 2$ and $q(A) = 1$.

In this case

$$A_{.3} = 3A_{.1} - 2A_{.2} ,$$

hence

$$\begin{bmatrix} 1 & 1 & 1 \\ -2 & 0 & -6 \\ 0 & 1 & -2 \end{bmatrix} = \begin{bmatrix} 1 & 1 \\ -2 & 0 \\ 0 & 1 \end{bmatrix} \begin{bmatrix} 1 & 0 & 3 \\ 0 & 1 & -2 \end{bmatrix}.$$

In the general case when $A(m \times n)$ has rank $r(A) = r$ which is such that

$$r < \min (m, n)$$

r columns of A are linearly independent.

By the above method, we define a matrix B, identical to A except for $(n - r)$ columns which are made zero. We next construct an appropriate E-matrix, so that

$$BE = A.$$

Now delete the zero columns of B and the corresponding rows of E to obtain the required product.

Example 1.1

Express each of the following matrices as a product:

(1) $$A = \begin{bmatrix} 1 & 1 & 3 & 2 \\ 2 & 0 & 2 & 2 \\ 3 & -1 & 1 & 2 \end{bmatrix}$$

(2) $$A = \begin{bmatrix} -1 & 1 & -2 & -3 \\ 2 & -2 & 4 & 6 \\ 1 & -1 & 2 & 3 \\ 0 & 0 & 0 & 0 \end{bmatrix}$$

Solution

(1) $\mathrm{r}(A) = 2$ and $\mathrm{q}(A) = 1$

$A_{\cdot 1}$ and $A_{\cdot 2}$ are linearly independent

$A_{\cdot 3} = A_{\cdot 1} + 2A_{\cdot 2}$ and $A_{\cdot 4} = A_{\cdot 1} + A_{\cdot 2}$

Hence

$$A = \begin{bmatrix} 1 & 1 \\ 2 & 0 \\ 3 & -1 \end{bmatrix} \begin{bmatrix} 1 & 0 & 1 & 1 \\ 0 & 1 & 2 & 1 \end{bmatrix}.$$

Notice that we can choose a different set of linearly independent columns of A, for example $\{A_{\cdot 1}, A_{\cdot 3}\}$.

Then

$$A_{\cdot 2} = \frac{1}{2}(A_{\cdot 3} - A_{\cdot 1}) \text{ and } A_{\cdot 4} = \frac{1}{2}(A_{\cdot 3} + A_{\cdot 1}),$$

so that

$$A = \begin{bmatrix} 1 & 0 & 3 & 0 \\ 2 & 0 & 2 & 0 \\ 3 & 0 & 1 & 0 \end{bmatrix} \begin{bmatrix} 1 & -\frac{1}{2} & 0 & \frac{1}{2} \\ 0 & 1 & 0 & 0 \\ 0 & \frac{1}{2} & 1 & \frac{1}{2} \\ 0 & 0 & 0 & 1 \end{bmatrix}.$$

Deleting the two columns of zeros in B and the corresponding rows in E, we obtain

$$A = \begin{bmatrix} 1 & 3 \\ 2 & 2 \\ 3 & 1 \end{bmatrix} \begin{bmatrix} 1 & -\frac{1}{2} & 0 & \frac{1}{2} \\ 0 & \frac{1}{2} & 1 & \frac{1}{2} \end{bmatrix}.$$

(2) In this case r(A) = 1 and q(A) = 3.
Since $A_{.2} = -A_{.1}$, $A_{.3} = 2A_{.1}$ and $A_{.4} = 3A_{.1}$, using the above process, we obtain

$$A = \begin{bmatrix} -1 \\ 2 \\ 1 \\ 0 \end{bmatrix} [1 \quad -1 \quad 2 \quad 3].$$

1.5 DETERMINANTS

It is assumed that the reader is familiar with various aspects of the theory of determinants, in particular with the expansion of a determinant in terms of the first cofactors of a matrix and more generally with Laplace's expansion.

(1) Sums of determinants

If a column of a square matrix is the sum of columns say

$$A = \begin{bmatrix} a_1 & a_2 & a_3 + a_4 \\ b_1 & b_2 & b_3 + b_4 \\ c_1 & c_2 & c_3 + c_4 \end{bmatrix}$$

then the determinant of A is a sum of determinants

$$|A| \text{ or } \det A = \begin{vmatrix} a_1 & a_2 & a_3 \\ b_1 & b_1 & b_3 \\ c_1 & c_2 & c_3 \end{vmatrix} + \begin{vmatrix} a_1 & a_2 & a_4 \\ b_1 & b_2 & b_4 \\ c_1 & c_2 & c_4 \end{vmatrix}.$$

For example, given

$$A = \begin{bmatrix} a_1 & a_2 & a_3 \\ b_1 & b_2 & b_3 \\ c_1 & c_2 & c_3 \end{bmatrix}$$

we can evaluate

$$C(\lambda) = \det [\lambda I - A]$$

(where I is the unit matrix and λ is a scalar) in terms of the principal minors of A.

If $\mathbf{e}_1$, $\mathbf{e}_2$ and $\mathbf{e}_3$ are elementary vectors defined in Section 1.1, and $A_{.1}$, $A_{.2}$ and $A_{.3}$ are the columns of A, we have

$$\det [\lambda I - A] = \det [\lambda \mathbf{e}_1 - A_{.1} \;\; \lambda \mathbf{e}_2 - A_{.2} \;\; \lambda \mathbf{e}_3 - A_{.3}].$$

By repeated application of the above result $\det [\lambda I - A]$ is seen to be the sum of 2^3 determinants;

$$\begin{aligned}\det [\lambda I - A] = {} & \det [\lambda \mathbf{e}_1 \; \lambda \mathbf{e}_2 \; \lambda \mathbf{e}_3] + \det [\lambda \mathbf{e}_1 \;\; \lambda \mathbf{e}_2 \; - A_{.3}] \\ & + \det [\lambda \mathbf{e}_1 \; - A_{.2} \; - A_{.3}] + \det [-A_{.1} \; \lambda \mathbf{e}_2 \; \lambda \mathbf{e}_3] \\ & + \det [\lambda \mathbf{e}_1 \; - A_{.2} \; - A_{.3}] + \det [-A_{.1} \;\; \lambda \mathbf{e}_2 \; - A_{.3}] \\ & + \det [-A_{.1} \; - A_{.2} \; \lambda \mathbf{e}_3] \\ & + \det [-A_{.1} \; - A_{.2} \; - A_{.3}] \\ = {} & \lambda^3 \det I - \lambda^2 \{\det [\mathbf{e}_1 \; \mathbf{e}_2 \; A_{.3}] \\ & + \det [\mathbf{e}_1 \; A_{.2} \; \mathbf{e}_3] + \det [A_{.1} \; \mathbf{e}_2 \; \mathbf{e}_3]\} \\ & + \lambda \{\det [\mathbf{e}_1 A_{.2} \; A_{.3}] + \det [A_{.1} \; \mathbf{e}_2 A_{.3}] \\ & + \det [A_{.1} \; A_{.2} \; \mathbf{e}_3]\} - \det [A_{.1} \; A_{.2} \; A_{.3}]\end{aligned}$$

where $I = [\mathbf{e}_1 \; \mathbf{e}_2 \; \mathbf{e}_3]$ is the unit matrix of order 3.

Each term of the coefficient of $(-\lambda^2)$ is a principal minor of order (3–2), that is a_{11}, a_{22} and a_{33} respectively.

Each term of the coefficient of λ is a principal minor of order (3–1). The term independent of λ, is the principal minor of order (3–0), that is the determinant of the matrix.

In the general case, when A is of order n, the coefficient of $(-1)^r \lambda^{n-r}$ in $C(\lambda)$ is the sum of all principal minors of A of order r.

(2) The Determinant of a Matrix Product

With the help of elementary matrices or by other means it is not difficult to show that if A and B are square matrices, then

$$\det AB = \det A \, . \det B.$$

(3) The Determinant of a Block Diagonal Matrix

If A is a block diagonal matrix

$$A = \begin{bmatrix} A_1 & 0 & 0 \ldots & 0 \\ 0 & A_2 & 0 \ldots & 0 \\ \vdots & \vdots & \vdots & \vdots \\ 0 & 0 & 0 & A_n \end{bmatrix}$$

where $A_i(i = \quad 2, \ldots, n)$ are square matrices then

$$\det A = \det A_1 \det A_2 \ldots \det A_n.$$

(4) The determinant of a Triangular Matrix

We consider an (upper) triangular matrix

$$A = \begin{vmatrix} a_{11} & a_{12} & a_{13} & \ldots & a_{1n} \\ 0 & a_{22} & a_{23} & \ldots & a_{2n} \\ 0 & 0 & a_{33} & \ldots & a_{3n} \\ \vdots & & & & \\ 0 & 0 & 0 & \ldots & a_{nn} \end{vmatrix}.$$

If A_{ij} is the cofactor of a_{ij} in A, then

$$|A| = \sum a_{ij} A_i$$

$$= a_{11} \begin{vmatrix} a_{22} & a_{23} & \ldots & a_{2n} \\ 0 & a_{33} & \ldots & a_{3n} \\ \vdots & \vdots & & \vdots \\ 0 & 0 & & a_{nn} \end{vmatrix} = \ldots = a_{11}a_{22}a_{33} \ldots a_{nn}.$$

1.6 THE DERIVATIVE OF A DETERMINANT

Let $A = [a_{ij}]$ where the entries $a_{ij} = a_{ij}(\lambda)$ $(i, j = 1, 2, \ldots, n)$ are differentiable functions of λ.

The derivatives of $|A|$ that is $d|A|/d\lambda$ is the sum of n determinants in which the jth determinant is identical to $|A|$ except for the entries of the jth column $(j = 1, 2, \ldots, n)$ which are replaced by the derivatives of the elements, that is by

$$[\dot{a}_{1j}\, \dot{a}_{2j} \ldots \dot{a}_{nj}]'$$

where

$$\dot{a}_{kj} = \frac{d}{d\lambda}(a_{kj}) \qquad (k = 1, 2, \ldots, n).$$

The result can be proved by expanding the determinant and then differentiating with respect to λ.

For example, if

$$A = \begin{bmatrix} a_{11} & a_{12} \\ a_{21} & a_{22} \end{bmatrix}$$

then

$$|A| = a_{11}a_{22} - a_{12}a_{21}$$

so that

$$\frac{\mathrm{d}|A|}{\mathrm{d}\lambda} = (\dot{a}_{11}a_{22} + a_{11}\dot{a}_{22}) - (\dot{a}_{12}a_{21} + a_{12}\dot{a}_{21})$$

$$= (\dot{a}_{11}a_{22} - a_{12}\dot{a}_{21}) + (a_{11}\dot{a}_{22} - \dot{a}_{12}a_{21})$$

$$= \begin{vmatrix} \dot{a}_{11} & a_{12} \\ \dot{a}_{21} & a_{22} \end{vmatrix} + \begin{vmatrix} a_{11} & \dot{a}_{12} \\ a_{21} & \dot{a}_{22} \end{vmatrix} .$$

A similar definition applies to higher derivatives of $|A|$, so that

$$\frac{\mathrm{d}^2|A|}{\mathrm{d}\lambda^2} = \frac{\mathrm{d}}{\mathrm{d}\lambda}\frac{\mathrm{d}|A|}{\mathrm{d}\lambda}$$

and in general

$$\frac{\mathrm{d}^r|A|}{\mathrm{d}\lambda^r} = \frac{\mathrm{d}}{\mathrm{d}\lambda}\frac{\mathrm{d}^{r-1}|A|}{\mathrm{d}\lambda^{r-1}} .$$

1.7 THE CHARACTERISTIC EQUATION

Given a square matrix $A = [a_{ij}]$ of order n, the matrix

$$g(\lambda) = [\lambda I - A] \tag{1.5}$$

where λ is a scalar is known as the *characteristic matrix* of A.

The determinant

$$C(\lambda) = \det\,[\lambda I - A] = \begin{vmatrix} \lambda - a_{11} & -a_{12} & -a_{13} & \dots -a_{1n} \\ -a_{21} & \lambda - a_{22} & -a_{23} & \dots -a_{2n} \\ -a_{31} & -a_{32} & \lambda - a_{33} & -a_{3n} \\ \vdots & \vdots & \vdots & \vdots \\ -a_{n1} & -a_{n2} & -a_{n3} & \dots \lambda - a_{nn} \end{vmatrix} \tag{1.6}$$

is called the *characteristic function* or the *characteristic polynomial* of A and the equation

$$C(\lambda) = 0 \tag{1.7}$$

is called the *characteristic equation* of A.

If $\mathbf{x} \neq \mathbf{0}$, then the scalars λ satisfying

$$[\lambda I - A]\mathbf{x} = \mathbf{0}$$

that is

$$g(\lambda)\mathbf{x} = \mathbf{0} \tag{1.8}$$

where the roots $\lambda_1, \lambda_2, \ldots, \lambda_n$ of $C(\lambda) = 0$ are called the *eigenvalues* (or the *characteristic* or *latent* values) of A and the corresponding vectors $\mathbf{x}$ in equation (1.8) are the *eigenvectors* (or the *characteristic* or *latent* vectors) of A.

We can write the characteristic equation in the form

$$C(\lambda) = \lambda^n + b_{n-1}\lambda^{n-1} + b_{n-2}\lambda^{n-2} + \ldots + b_1\lambda + b_0 = 0 \qquad (1.9)$$

We have seen in Section 1.5 that b_{n-r}, the coefficient of λ^{n-r} in (1.9) is (except for the sign) the sum of all the principal minors A of order r.

In particular b_0 is (except for the sign) the principal minor of order n, that is the det A.

Indeed from (1.9)

$$b_0 = C(0) = \det\,[-A] \text{ from (1.6)}$$

Also b_{n-1} is (except for the sign) the sum of the principal minors of order 1, that is

$$\begin{aligned} b_{n-1} &= a_{11} + a_{22} : \ldots + a_{nn} \\ &= \operatorname{tr} A \end{aligned} \qquad (1.10)$$

For definition of tr A and its properties see Section 1.11.

We can also write $C(\lambda)$ as

$$C(\lambda) = (\lambda - \lambda_1)(\lambda - \lambda_2) \ldots (\lambda - \lambda_n) \qquad (1.11)$$

where $\lambda_1, \lambda_2, \ldots, \lambda_n$ are the characteristic roots of A.

If the eigenvalues $\lambda_1, \lambda_2, \ldots, \lambda_n$ are all distinct, that is

$$\lambda_i \neq \lambda_j \quad \text{whenever } i \neq j$$

we say that the eigenvalues are *simple*. On the other hand, if r of the eigenvalues are equal, say $\lambda_1 = \lambda_2 = \ldots = \lambda_r$, then

$$(\lambda - \lambda_1)^r$$

is a factor of $C(\lambda)$, and we say that $\lambda = \lambda_1$ is a eigenvalue of *multiplicity r*.

Another useful concept refers to the magnitude of the largest eigenvalue of a matrix.

If $\lambda_1, \lambda_2, \ldots, \lambda_n$ are the eigenvalues of $A(n \times n)$, then

$$\rho = \max_i |\lambda_i| \qquad (i = 1, \ldots, n)$$

is called the *spectral radius* of A, which we will sometimes denote by $\rho(A)$. The set $\{\lambda_1, \lambda_2, \ldots, \lambda_n\}$ is called the *spectrum* of A.

There are numerous other important results to be found in literature (see [B7] for example) referring to the eigenvalues of a matrix.

For example (I) if A has eigenvalues $\lambda_1, \lambda_2, \ldots, \lambda_n$ then A^r has eigenvalues

$$\lambda_1^r, \lambda_2^r, \ldots, \lambda_n^r.$$

and (II) if λ is an eigenvalue of A of multiplicity k, then λ^r is an eigenvalue of A^r of multiplicity at least k.

We find from (1.9) and (1.11) that

$$b_0 = C(0) = \det(-A) = (-1)^n \lambda_1 \lambda_2 \ldots \lambda_n.$$

Since $\det(-A) = (-1)^n \det A$, it follows that

$$|A| = \lambda_1 \lambda_2 \ldots \lambda_n. \tag{1.12}$$

Bounds for the Spectral Radius of a Matrix

One of the most useful theorems giving the bounds for the eigenvalues of a matrix A (A may be either complex or real) is known as the *Gershgorin theorem.*

Theorem

Let $A(n \times n) = [a_{ij}], P_i = \sum_{\substack{j=1 \\ j \neq i}}^{n} |a_{ij}| \qquad (i = 1, 2, \ldots, n)$

then every eigenvalue λ of A satisfies at least one of the inequalities:

$$|\lambda - a_{ii}| \leqslant P_i \qquad (i = 1, 2, \ldots, n).$$

Notice that in the complex plane

$$|\lambda - a| \leqslant P.$$

represents a disc centred at $\lambda = a$ having a radius P. So the above theorem can be stated as follows:

All eigenvalues λ of A lie in the union of discs

$$|\lambda - a_{ii}| \leqslant P_i$$

Proof

Let $\mathbf{z}$ be an eigenvector corresponding to the eigenvalue λ of A, then

$$[\lambda I - A]\, \mathbf{z} = \mathbf{0} \quad \text{where} \quad \mathbf{z} = [z_1, z_2, \ldots, z_n]'.$$

We can choose $\mathbf{z}$ in a normalised form, such that

$$|z_r| = \max \; |z_1|, |z_2|, \ldots, |z_n| \; = 1$$

On equating the rth elements of the equation $[\lambda I - A]\mathbf{z} = \mathbf{0}$, we obtain

$$(\lambda - a_{rr})z_r + \sum_{j \neq r} (-a_{rj})z_j = 0$$

so that

$$|\lambda - a_{rr}| = \left| \sum_{j \neq r} a_{rj} z_j \right|$$

$$\leqslant \sum_{j \neq r} |a_{rj}| \, |z_j| \leqslant \sum_{j \neq r} |a_{rj}|$$

(Since $|z_j| \leqslant 1$ $(j \neq r)$).

So λ lies in one of the above discs.

The result now follows.

Corollary

If $Q_j = \sum_{\substack{i=1 \\ i \neq j}}^{n} |a_{ij}|$, then all eigenvalues λ of A lies in the union of discs

$$|\lambda - a_{jj}| \leqslant Q_j \qquad (j = 1, 2, \ldots, n).$$

Proof

Apply the above theorem to the matrix A'. ■

For example, let

$$A = \begin{bmatrix} 1 & 1.5 & -0.5 \\ 1 & 0 & -1 \\ i & 1 & 5i \end{bmatrix}.$$

The three discs within which all eigenvalues of A are to be found are shown in Fig. 1. The three discs form two *disjoint components* D_1 and D_2. By a continuity argument (some indication of this argument is given in Section 4.5.2) it can be shown that D_1 contains 1 eigenvalue of A and D_2 contains 2 eigenvalues of A.

Exercise 1

The matrix $A = [a_{ij}]$ is such that

$$|a_{ii}| > P_i \qquad (i = 1, 2, \ldots, n)$$

(we call such matrices strictly diagonally dominant, see Section 5.2 where a more general definition is given).

Let

$$d_i = |a_{ii}| - P_i \qquad (i = 1, 2, \ldots, n)$$

show that

$$|A| \geqslant d_1 d_2 \ldots d_n$$

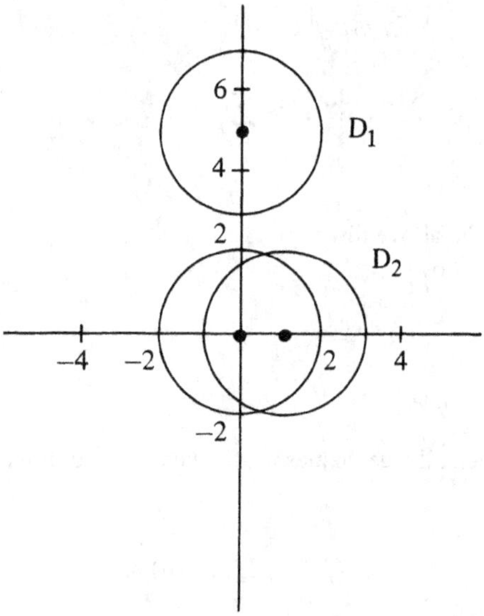

Fig. 1.

Solution

Let $B = [b_{ij}]$

where $$b_{ij} = \frac{a_{ij}}{d_i}.$$

Then $|A| = d_1 d_2 \ldots d_n \; |B| = (d_1 d_2 \ldots d_n)(\lambda_1 \lambda_2 \ldots \lambda_n)$ where λ_i $(i = 1, 2, \ldots, n)$ are the eigenvalues of B. If λ is any eigenvalue of B, then

$$|\lambda - b_{ii}| \leqslant S_i \text{ for some } i$$

where

$$S_i = \sum_{\substack{j=1 \\ j \neq i}}^{n} |b_{ij}|$$

But $|\lambda - b_{ii}| \geqslant ||\lambda| - |b_{ii}||$ so that $S_i \geqslant |b_{ii}| - |\lambda|$ or

$$|\lambda| \geqslant |b_{ii}| - S_i = \frac{|a_{ii}| - P_i}{d_i} = \frac{d_i}{d_i} = 1$$

where

$$P_i = \sum_{\substack{j=1 \\ j \neq i}}^{n} a_{ij}.$$

So each eigenvalue λ_i of B is such that

$$|\lambda_i| \geqslant 1$$

Hence $|A| \geqslant d_1 d_2 \ldots d_n$.

Exercise 2

If a matrix $A = [a_{ij}]$ is strictly diagonally dominant and has all diagonal entries positive and λ is an eigenvalue of A, show that

$$\text{Re}\,\lambda > 0.$$

Solution

All the eigenvalues λ_i of A lie in the union of the discs

$$|\lambda - a_{ii}| \leqslant P_i.$$

Since $a_{ii} > P_i$ $(i = 1, 2, \ldots, n)$ all the discs are contained in the positive part of the complex plane.

Hence every point of the union of the discs has a positive real part. The conclusion follows.

1.8 THE ADJOINT OF THE CHARACTERISTIC MATRIX

Using the notation of the previous section, $g(\lambda) = [\lambda I - A] = [g_{ij}]$ is the characteristic matrix of A and $G(\lambda) = [G_{ij}]'$ is the *adjoint matrix* of g, where G_{ij} is the cofactor of g_{ij} in g and the dash refers to the transpose of the matrix.

We write the adjoint matrix of g either as G or as adj g.

In the text other notation is sometimes used. For example $B(\lambda)$ is sometimes used to denote the matrix adj $[\lambda I - A]$.

The fundamental relations associated a matrix $[\lambda I - A]$ and its adjoint are

$$\begin{aligned} g \text{ adj } g &= |g|\, I \\ &= \mathrm{C}(\lambda)\, I \text{ using } (1.6) \end{aligned} \tag{1.13}$$

and

$$|\text{adj } g| = |g|^{n-1} \tag{1.14}$$

which follows from (1.13).

Since the right-hand side of (1.13) is a diagonal matrix, the matrices on the left-hand side commute, and we have

$$g(\lambda)\, G(\lambda) = G(\lambda)\, g(\lambda) = \mathrm{C}(\lambda) I \tag{1.15}$$

Differentiating (1.15) r times with respect to λ, we obtain

$$\frac{\mathrm{d}^r}{\mathrm{d}\lambda^r}\{g(\lambda)\, G(\lambda)\} = \frac{\mathrm{d}^r}{\mathrm{d}\lambda^r}\{G(\lambda)\, g(\lambda)\} = \frac{\mathrm{d}^r \mathrm{C}(\lambda)}{\mathrm{d}\lambda^r} I. \tag{1.16}$$

We will be interested in the form of $G(\lambda)$ for $\lambda = \lambda_j$ where λ_j is an eigenvalue of A. For this purpose we consider the characteristic equation in the form (1.11). On differentiation with respect to λ, we obtain

$$C'(\lambda) = \sum_{i=1}^{n} \frac{C(\lambda)}{\lambda - \lambda_i}. \tag{1.17}$$

(1.17) is true for a simple root $\lambda = \lambda_j$. It is also true for a multiple root if in the summation (1.17) a term $C(\lambda)/\lambda - \lambda_j$ is counted r times whenever $\lambda = \lambda_j$ has multiplicity r.

We obtain the following important results, which follow immediately from (1.17).

(1) If $\lambda = \lambda_j$ is a simple eigenvalue of A then $C'(\lambda_j) \neq 0$.
(2) If $\lambda = \lambda_j$ is an eigenvalue of multiplicity $r > 1$, then $C'(\lambda_j) = 0$.

We now consider the form of the adjoint matrix $G(\lambda)$ at $\lambda = \lambda_j$, when A is a matrix of order n. First we assume that $\lambda = \lambda_j$ is a *simple eigenvalue*. From the above discussion

$$C'(\lambda_j) \neq 0.$$

But using the technique of determinant differentiation (discussed in Section 1.6) on the characteristic function $C(\lambda)$ defined by equation (1.6) we see that $C'(\lambda_j)$ is a linear combination of the principal minors of $g(\lambda_j)$. It follows that at least one first minor of $g(\lambda_j)$ is not zero. But (see 1.8)), $C(\lambda_j) = \det g(\lambda_j) = 0$ hence

$$\text{rank}\,\{g(\lambda_j)\} = n - 1.$$

By (1.15), since $C(\lambda_j) = 0$,

$$g(\lambda_j)\, G(\lambda_j) = 0. \tag{1.18}$$

Now apply (1.4), the rank theorem for a matrix product, we find

$$\text{rank}\,\{G(\lambda_j)\} = 1.$$

Finally, by the method explained in Section 1.4 (see Example 1.1),

$$G(\lambda_j) = \mathbf{x} \,.\, \mathbf{y}' \tag{1.19}$$

where $\mathbf{x}$ and $\mathbf{y}$ are vectors of order n.

Next we assume that $\lambda = \lambda_j$ is an eigenvalue of A of multiplicity $r > 1$. This time

$$C'(\lambda_j) = 0.$$

It follows that all the first minors of $g(\lambda_j)$ are zero. Hence

$$\text{rank}\,\{g(\lambda_j)\} < n - 1.$$

Since

$$g(\lambda_j)\, G(\lambda_j) = 0,$$

on application of the rank theorem (1.14), we find

$$\mathrm{r}\,\{G(\lambda_j)\} = 0$$

which implies that

$$G(\lambda_j) = 0. \tag{1.20}$$

Example 1.2
Verify the results established in this section for the matrices

$$(1)\quad A = \begin{bmatrix} 3 & -1 & 0 \\ -1 & 2 & -1 \\ 0 & -1 & 3 \end{bmatrix} \quad \text{and (2)}\quad A = \begin{bmatrix} 2 & 2 & 1 \\ 1 & 3 & 1 \\ 2 & 2 & 2 \end{bmatrix}$$

Solution

(1)

$$g(\lambda) = \begin{bmatrix} \lambda-3 & 1 & 0 \\ 1 & \lambda-2 & 1 \\ 0 & 1 & \lambda-3 \end{bmatrix}.$$

so that the characteristic equation is

$$C(\lambda) = (\lambda-1)(\lambda-3)(\lambda-4).$$

In this case, $\lambda_1 = 1$, $\lambda_2 = 3$ and $\lambda_3 = 4$ are simple eigenvalues of $g(\lambda)$. The adjoint matrix of $g(\lambda)$ is found to be

$$G(\lambda) = \begin{bmatrix} \lambda^2-5\lambda+5 & 3-\lambda & 1 \\ 3-\lambda & (3-\lambda)^2 & 3-\lambda \\ 1 & 3-\lambda & \lambda^2-5\lambda+5 \end{bmatrix}.$$

Matrix multiplication shows that

$$g(\lambda)\, G(\lambda) = G(\lambda)\, g(\lambda) = C(\lambda)\, I.$$

Also

$$C'(\lambda) = (\lambda-1)(\lambda-3)(\lambda-4)\left[\frac{1}{\lambda-1} + \frac{1}{\lambda-3} + \frac{1}{\lambda-4}\right].$$

By the method of Section 1.4, we find

$$G(\lambda_1) = \begin{bmatrix} 1 & 2 & 1 \\ 2 & 4 & 2 \\ 1 & 2 & 1 \end{bmatrix} = \begin{bmatrix} 1 \\ 2 \\ 1 \end{bmatrix} [1 \;\; 2 \;\; 1].$$

Similarly,

$$G(\lambda_2) = \begin{bmatrix} 1 \\ 0 \\ -1 \end{bmatrix} [-1 \;\; 0 \;\; 1] \quad \text{and} \quad G(\lambda_3) = \begin{bmatrix} 1 \\ -1 \\ 1 \end{bmatrix} [1 \;\; -1 \;\; 1]$$

(2)

$$g(\lambda) = \begin{bmatrix} \lambda - 2 & -2 & -1 \\ -1 & \lambda - 3 & -1 \\ -1 & -2 & \lambda - 2 \end{bmatrix}$$

and

$$C(\lambda) = (\lambda - 1)^2 (\lambda - 5).$$

This time $\lambda_1 = 1$ is an eigenvalue of A multiplicity 2. $\lambda_2 = 5$ is a simple eigenvalue.

$$G(\lambda) = \begin{bmatrix} (\lambda - 1)(\lambda - 4) & 2(\lambda - 1) & \lambda - 1 \\ \lambda - 1 & (\lambda - 1)(\lambda - 3) & \lambda - 1 \\ \lambda - 1 & 2(\lambda - 1) & (\lambda - 1)(\lambda - 4) \end{bmatrix}.$$

Again, we find

$$g(\lambda)\, G(\lambda) = G(\lambda)\, g(\lambda) = C(\lambda) I.$$

$$C'(\lambda) = (\lambda - 1)^2 (\lambda - 5) \left[\frac{2}{\lambda - 1} + \frac{1}{\lambda - 5} \right].$$

$$G(\lambda_1) = \begin{bmatrix} 0 & 0 & 0 \\ 0 & 0 & 0 \\ 0 & 0 & 0 \end{bmatrix}$$

as predicted by (1.20), and

$$G(\lambda_2) = \begin{bmatrix} 4 & 8 & 4 \\ 4 & 8 & 4 \\ 4 & 8 & 4 \end{bmatrix} = \begin{bmatrix} 4 \\ 4 \\ 4 \end{bmatrix} [1 \;\; 2 \;\; 1].$$

We next consider another result which will be found useful.

Let λ_j be a simple eigenvalue and $\mathbf{x}_j$ the corresponding eigenvector of a matrix A or order n.

By (1.8)

$$g(\lambda_j)\mathbf{x}_j = \mathbf{o} \tag{1.21}$$

Let **x** by any non-zero column of $G(\lambda_j)$, then by (1.19) (also see Example 1.2)

$$G(\lambda_j) = \mathbf{x} \,.\, \mathbf{y}'$$

where

$$\mathbf{y}' \neq \mathbf{o}.$$

Hence from (1.18)

$$g(\lambda_j)\mathbf{x}\,\mathbf{y}' = 0.$$

it follows that

$$g(\lambda_j)\mathbf{x} = \mathbf{o}. \tag{1.22}$$

On comparing (1.21) and (1.22) we conclude that $\mathbf{x}_j$, an eigenvector of A corresponding to a simple eigenvalue λ_j, is proportional (or equal) to **x**, any non-zero column of $G(\lambda_j)$. Conversely, if $\lambda = \lambda_j$ is a simple eigenvalue, each column of $G(\lambda_j)$ is either proportional to the eigenvector $\mathbf{x}_j$ or is a column of zeros.

Example 1.3

Illustrate the above result for the matrices considered in Example 1.2.

Solution

(1) All the eigenvalues for this matrix are simple. The eigenvalues and the corresponding eigenvectors are

$$\lambda_1 = 1 \quad \mathbf{x}_1 = [1 \quad 2 \quad 1]'$$

$$\lambda_2 = 3 \quad \mathbf{x}_2 = [1 \quad 0 \; -1]'$$

$$\lambda_3 = 4 \quad \mathbf{x}_3 = [1 \; -1 \quad 1]'.$$

We have found in Example 1.2 that the first columns of $G(\lambda_1)$, $G(\lambda_2)$ and $G(\lambda_3)$ are respectively equal to the eigenvectors $\mathbf{x}_1$, $\mathbf{x}_2$ and $\mathbf{x}_3$.

(2) In this case only the eigenvalue $\lambda_2 = 5$ is simple. The corresponding eigenvector is $\mathbf{x}_2 = [1\ 1\ 1]'$, it is proportional to the first column of $G(\lambda_2)$.

1.9 SPECTRAL DECOMPOSITION

In this section we consider matrices which can be transformed into diagonal forms. Such matrices are called *nondefective*. In particular a matrix whose eigenvalues are all distinct is nondefective. Of course matrices with multiple eigenvalues can also be nondefective.

Theorem (Spectral Decomposition Theorem)
Let $A(n \times n)$ be a nondefective matrix having distinct eigenvalues $\lambda_1, \lambda_2, \ldots, \lambda_k$ where λ_j occurs with multiplicity n_j $(j = 1, 2, \ldots, k)$, so that

$$n_1 + n_2 + \ldots + n_k = n.$$

Then

$$A = \lambda_1 P_1 + \lambda_2 P_2 + \ldots + \lambda_k P_k,$$

where P_j is of order $(n \times n)$, and

(1) $P_j^2 = P_j$ (P_j is idempotent)

(2) $P_i P_j = 0$ $(i \neq j)$ (P_j are 'mutually orthogonal')

(3) $P_1 + P_2 + \ldots + P_k = I$

(4) $P_j A = A P_j$ (P_j commutes with A)

for $i, j = 1, 2, \ldots, k$.

Proof
Since A is nondefective, there exists a matrix Q such that

$$Q^{-1} A Q = \text{diag}\{\lambda_1, \lambda_2, \ldots, \lambda_k\} = \Lambda \text{ (say)}$$

We can write Λ as

$$\Lambda = \lambda_1 E_1 + \lambda_2 E_2 + \ldots + \lambda_k E_k$$

where E_j is a diagonal matrix with unity in the positions where λ_j occurs in Λ and zeros elsewhere.

It follows that

$$E_j^2 = E_j \quad (j = 1, 2, \ldots, k)$$

and

$$E_i E_j = 0 \text{ since } \lambda_j \text{ and } \lambda_i \text{ are distinct.}$$

Also,

$$\sum_{j=1}^{k} E_j = I.$$

Since

$$A = Q \Lambda Q^{-1},$$

it follows that

$$\begin{aligned} A &= \lambda_1 Q E_1 Q^{-1} + \lambda_2 Q E_2 Q^{-1} + \ldots + \lambda_k Q E_k Q^{-1} \\ &= \lambda_1 P_1 + \lambda_2 P_2 + \ldots + \lambda_k P_k \end{aligned}$$

where

$$P_j = QE_jQ^{-1} \qquad (j = 1, 2, \ldots, k)$$

Since E_j satisfies (1), (2) and (3), so does $P_j.(j = 1, 2, \ldots, k)$. Notice that by (1) and (2)

$$AP_j = \lambda_j P_j.$$

Similarly

$$P_j A = \lambda_j P_j \qquad (j = 1, 2, \ldots, k)$$

Condition (4) follows.

Example 1.4
Consider

$$A = \begin{bmatrix} 2 & 1 & 1 \\ 1 & 2 & 1 \\ 0 & 0 & 1 \end{bmatrix}.$$

The eigenvalues of A are $\lambda_1 = 1, \lambda_2 = 1$ and $\lambda_3 = 3$.

$$Q = \begin{bmatrix} 1 & 0 & 1 \\ 0 & 1 & 1 \\ -1 & -1 & 0 \end{bmatrix} \text{ and } Q^{-1} = \frac{1}{2}\begin{bmatrix} 1 & -1 & -1 \\ -1 & 1 & -1 \\ 1 & 1 & 1 \end{bmatrix}.$$

Also

$$E_1 = \operatorname{diag}\{1, 1, 0\} \text{ and } E_2 = \operatorname{diag}\{0, 0, 1\}$$

so that

$$P_1 = \begin{bmatrix} \frac{1}{2} & -\frac{1}{2} & -\frac{1}{2} \\ -\frac{1}{2} & \frac{1}{2} & -\frac{1}{2} \\ 0 & 0 & 1 \end{bmatrix} \text{ and } P_2 = \begin{bmatrix} \frac{1}{2} & \frac{1}{2} & \frac{1}{2} \\ \frac{1}{2} & \frac{1}{2} & \frac{1}{2} \\ 0 & 0 & 0 \end{bmatrix}. \quad \blacksquare$$

We generalise the above theorem in the following corollary.

Corollary
If A is nondefective having k distinct eigenvalues $\lambda_1, \lambda_2, \ldots, \lambda_k$ and $f(x)$ is a polynomial, then

$$f(A) = f(\lambda_1)P_1 + g(\lambda_2)P_2 + \ldots + f(\lambda_k)P_k.$$

Proof
Since

$$A = \lambda_1 P_1 + \lambda_2 P_2 + \ldots + \lambda_k P_k$$

then

$$A^r = \lambda_1^r P_1 + \lambda_2^r P_2 + \ldots + \lambda_k^r P_k \qquad (r = 1, 2, \ldots)$$

by conditions (1) and (2) of the theorem.
The result follows.
In particular notice that

$$e^A = e^{\lambda_1} P_1 + e^{\lambda_2} P_2 + \ldots + e^{\lambda_k} P_k.$$

Another useful associated result is the following.
Since

$$A\mathbf{x} = \lambda \mathbf{x}$$

where λ and $\mathbf{x}$ are an eigenvalue and the corresponding eigenvector of A, it follows that

$$A^r \mathbf{x} = \lambda^r \mathbf{x} \qquad (r = 1, 2, \ldots)$$

so that if $f(x)$ is a polynomial

$$f(A)\mathbf{x} = f(\lambda)\mathbf{x}.$$

Assuming that $A(n \times n)$ has eigenvalues λ_i and corresponding eigenvectors $\mathbf{x}_i$ $(i = 1, 2, \ldots, n)$ then

$$e^A \mathbf{x}_i = e^{\lambda_i} \mathbf{x}_i.$$

Hence

$$e^A Q = Q e^{\Lambda}$$

where

$$Q \text{ is the modal matrix } [\mathbf{x}_1, \mathbf{x}_2, \ldots, \mathbf{x}_n]$$

and

$$\Lambda = \text{diag}\{\lambda_1, \lambda_2, \ldots, \lambda_n\}$$

It follows that

$$e^A = Q e^{\Lambda} Q^{-1}.$$

This last equation also implies that

$$e^{Q\Lambda Q^{-1}} = Q e^{\Lambda} Q^{-1}.$$

and conversely

$$Q^{-1} e^A Q = e^{Q^{-1}AQ} = e^{\Lambda}.$$

1.10 INNER PRODUCT AND NORMS

For many purposes it is useful to have a measure of the size of a vector or a matrix, we refer to such a measure as the *norm* of the vector (or the matrix). We have here an analogy with statistics, where the mean of a sample is a useful measure for the set.

Generally, we denote the norm of a vector $\mathbf{x}$ by $\|\mathbf{x}\|$. Since the norm is an indication of the 'size' or the 'length' of a vector it is desirable that it should satisfy the following properties:

(a) $\|\mathbf{x}\| \geqslant o$ (with $\|\mathbf{x}\| = o$ for $\mathbf{x} = \mathbf{o}$)

(b) $\|\lambda\mathbf{x}\| = |\lambda|\,\|\mathbf{x}\|$ (λ is a scalar, real or complex) (1.23)

(c) $\|\mathbf{x}+\mathbf{y}\| \leqslant \|\mathbf{x}\| + \|\mathbf{y}\|$ (the triangle inequality).

There are many possible norms satisfying these conditions. For example, the *Euclidean* norm

$$\|\mathbf{x}\| = \left\{ \sum_{i=1}^{n} |x_i|^2 \right\}^{1/2} \tag{1.24}$$

where

$$\mathbf{x}' = [x_1 x_2 \ldots x_n].$$

To prove various properties of norms, we make use of the *inner product* (or the *scalar product*) of two vectors $\mathbf{x}$ and $\mathbf{y}$ of order n having real or complex elements $\{x_i\}$ and $\{y_i\}$ respectively.

It is denoted by $(\mathbf{x}, \mathbf{y})$ and defined as

$$(\mathbf{x}, \mathbf{y}) = \mathbf{x}^*\mathbf{y} = \bar{x}_1 y_1 + \bar{x}_2 y_2 + \ldots + \bar{x}_n y_n \tag{1.25}$$

where $\bar{x}_j$ is the complex conjugate of x_j.

Notice that in the definition we are not distinguishing between the scalar $(\mathbf{x}, \mathbf{y})$ and the matrix $\mathbf{x}^*\mathbf{y}$ of order 1.

The inner product has the following properties:

(a) $(\mathbf{x}, \mathbf{y}) = \overline{(\mathbf{y}, \mathbf{x})}$

(the bar over an expression indicates its complex conjugate)

(b) $(\lambda\mathbf{x}, \mathbf{y}) = \bar{\lambda}(\mathbf{x}, \mathbf{y})$ (1.26)

(c) $(\mathbf{x}+\mathbf{y}, \mathbf{z}) = (\mathbf{x}, \mathbf{z}) + (\mathbf{y}, \mathbf{z})$.

(d) $(\mathbf{x}, \mathbf{x}) > o$ if $\mathbf{x} \neq \mathbf{o}$.

These properties are an immediate consequence of the defintion (1.25).

In terms of the inner product notation, the Euclidean norm (1.24) is written as

$$\|\mathbf{x}\| = (\mathbf{x}, \mathbf{x})^{1/2} \tag{1.27}$$

If $(\mathbf{x}, \mathbf{y}) = 0$, we say that the two vectors $\mathbf{x}$ and $\mathbf{y}$ are *orthogonal*.

If the vectors in the set $S = \{\mathbf{x}_1, \mathbf{x}_2, \ldots, \mathbf{x}_n\}$ are

(1) mutually orthogonal
(2) $(\mathbf{x}_i, \mathbf{x}_i) = 1 \; (i = 1, 2, \ldots, n)$

then we say that S is an *orthonormal* set of vectors.

The two conditions (1) and (2) above are frequently written in the form

$$(\mathbf{x}_i, \mathbf{x}_j) = \delta_{ij} \tag{1.28}$$

where δ_{ij} is the *Kronecker delta*,

$$\delta_{ij} = \begin{cases} 0 & \text{if } i \neq j \\ 1 & \text{if } i = j \end{cases}$$

An important property involving vector norms is the *Schwarz inequality*

$$|\mathbf{x}^*\mathbf{y}| \leqslant \|\mathbf{x}\| \, \|\mathbf{y}\| \tag{1.29}$$

which can be proved in the following manner.

Let λ be any (complex) scalar.

By properties (1.26)

$$\begin{aligned} 0 \leqslant \|\mathbf{x} + \lambda\mathbf{y}\|^2 &= (\mathbf{x} + \lambda\mathbf{y}, \mathbf{x} + \lambda\mathbf{y}) \\ &= (\mathbf{x} + \lambda\mathbf{y})^*(\mathbf{x} + \lambda\mathbf{y}) \\ &= \mathbf{x}^*\mathbf{x} + \lambda\mathbf{x}^*\mathbf{y} + \bar{\lambda} \quad \bar{\lambda}\mathbf{y}^*\mathbf{x} + \bar{\lambda}\lambda\mathbf{y}^*\mathbf{y}. \end{aligned}$$

Hence

$$0 \leqslant \|\mathbf{x}\|^2 + |\lambda|^2 \, \|\mathbf{y}\|^2 + \lambda\mathbf{x}^*\mathbf{y} + \bar{\lambda}\mathbf{y}^*\mathbf{x}.$$

If $\mathbf{x}^*\mathbf{y} = 0$, then (1.29) is (trivially) true.

So consider that $\mathbf{x}^*\mathbf{y} \neq 0$ and let

$$\lambda = -\frac{\|\mathbf{x}\|^2}{\mathbf{x}^*\mathbf{y}}$$

so that

$$\bar{\lambda} = -\frac{\|\mathbf{x}\|^2}{\mathbf{y}^*\mathbf{x}}$$

then the above becomes

$$0 \leqslant \frac{\|\mathbf{x}\|^4 \|\mathbf{y}\|^2}{|\mathbf{x}^*\mathbf{y}|^2} - \|\mathbf{x}\|^2$$

so that

$$\|\mathbf{x}\|^2 \|\mathbf{y}\| \geqslant \|\mathbf{x}\| \, |\mathbf{x}^*\mathbf{y}|$$

and (1.28) follows.

To prove that (1.24) does define a vector norm, we must show that it satisfies conditions (1.23).

Conditions (a) and (b) are obvious. The triangle inequality is more complicated to prove.

Example 1.5

Prove the triangle inequality for the Euclidean norm.

Solution

$$\begin{aligned}\|\mathbf{x}+\mathbf{y}\|^2 &= (\mathbf{x}+\mathbf{y},\mathbf{x}+\mathbf{y})\\ &= \|\mathbf{x}\|^2 + (\mathbf{x},\mathbf{y}) + (\mathbf{y},\mathbf{x}) + \|\mathbf{y}\|^2\\ &\leqslant \|\mathbf{x}\|^2 + 2\|\mathbf{x}\|\,\|\mathbf{y}\| + \|\mathbf{y}\|^2 \qquad \text{(using (1.29))}\\ &= (\|\mathbf{x}\| + \|\mathbf{y}\|)^2 .\end{aligned}$$

It follows that

$$\|\mathbf{x}+\mathbf{y}\| \leqslant \|\mathbf{x}\| + \|\mathbf{y}\|. \quad \blacksquare$$

There are of course other possible definitions of vector norms. For example

$$\|\mathbf{x}\|_1 = |x_1| + |x_2| + \ldots + |x_n| \tag{1.30}$$

or

$$\|\mathbf{x}\|_\infty = \max\left\{|x_1|,\ |x_2|, \ldots, |x_n|\right\} \tag{1.31}$$

Notice the suffixes 1 and ∞ used to identify the norms. Similarly, the Euclidean norm (1.24) is frequently suffixed by a '2', that is

$$\|x\|_2 .$$

There is no difficulty in proving that these norms satisfy conditions (1.23).

We next investigate a method of defining matrix norms from the corresponding vector norms – they are known as *induced matrix norms.*

If we have a matrix $A(m \times n)$ we can consider that it is representing a linear transformation

$$T : U \to V$$

(see [B7] p. 77) where U and V are vector spaces of orders $\boldsymbol{n}$ and $\boldsymbol{m}$ respectively. Let $\mathbf{x} \neq \mathbf{0} \in U$, and consider a norm

$$\|\mathbf{x}\|_U .$$

Similarly if $\mathbf{y} \in V$ then its norm is

$$\|\mathbf{y}\|_V .$$

The norms $\|\,.\,\|_U$ and $\|\,.\,\|_V$ can be different although in general they are chosen to be the same.

The interesting point is that $\|\mathbf{x}\|_U$ indicates the size of the vector in U, whereas $\|A\mathbf{x}\|_V$ indicates the size of this vector in V after it has been transformed by A.

Hence the ratio

$$\frac{\|A\mathbf{x}\|_V}{\|\mathbf{x}\|_U} \tag{1.32}$$

is the magnification of this vector as it is transformed from U to V and so is an indication of the 'size' of the matrix A.

Given a matrix $A(m \times n)$ (we will consider only finite dimensional matrices), it is seen that for each choice of the vector $\mathbf{x}$, we obtain a different value of the ratio (1.32). Nevertheless it can be shown that for every choice of $\mathbf{x} \neq \mathbf{0}$, this ratio has a supremum (the least upper bound) which is defined as the *norm of the matrix*, $\|A\|$, so

$$\|A\| = \sup_{\mathbf{x} \neq 0} \frac{\|A\mathbf{x}\|}{\|\mathbf{x}\|}.$$

Example 1.6

Let $A(m \times n) = [a_{ij}]$ where $a_{ij} \in C$ (complex numbers).

(1) If $\mathbf{z} \neq \mathbf{0}$, show that

$$\frac{\|A\mathbf{z}\|_\infty}{\|x\|_\infty} \leqslant \delta$$

where

$$\delta = \max_i \sum_{j=1}^{n} |a_{ij}| = \max |A_{i.}'|$$

($A_{i.}'$ is the ith row of A, see Section 1.2 for notation.)

(2) Find the vector $\mathbf{z}$ for which the equality holds.

Solution

(1) If $\mathbf{z} = [z_1 z_2 \ldots z_n]'$. Let

$$|z_k| = \max \{|z_1|, |z_2| \ldots |z_n|\}.$$
$$= \|\mathbf{z}\|_\infty.$$

We denote the ith component of the vector $A\mathbf{z}$ by $(A\mathbf{z})_i$ we have

$$\|A\mathbf{z}\|_\infty = \max_i |(A\mathbf{z})_i| = \max_i \left| \sum_{j=1}^{n} a_{ij} z_j \right| \leqslant \max_i \sum_{j=1}^{n} |a_{ij}|\,|z_j|$$
$$\leqslant \delta \|\mathbf{z}\|_\infty.$$

The results follows.

(2) Assume that for the matrix A,

$$\delta = \sum_{j=1}^{n} |a_{rj}|.$$

Let $a_{rj} = \alpha_j + i\beta_j$ and choose $z_j = u_j + iv_j$ so that $a_{rj}z_j = |a_{rj}|$, then

$$u_j = \frac{\alpha_j}{(\alpha_j^2 + \beta_j^2)^{1/2}} \quad \text{and} \quad v_j = \frac{-\beta_j}{(\alpha_j^2 + \beta_j^2)^{1/2}} \qquad (j = 1, 2, \ldots, n)$$

where all the α, β, u and v are real. We now have

$$\|A\mathbf{z}\|_\infty = \left|\sum_{j=1}^{n} a_{rj} z_j\right| = \left|\sum_{j=1}^{n} |a_{rj}|\right| = \delta.$$

Hence the required vector is $\mathbf{z} = [z_1 z_2 \ldots z_n]'$, for which the components $z_j = u_j + iv_j\,(j = 1, 2, \ldots, n)$ are related in the above specified manner to a_{rj}.

The above discussion leads us to define various matrix norms induced by the corresponding vector norms as

$$\|A\|_j = \sup_{\mathbf{x} \neq 0} \left\{ \frac{\|A\mathbf{x}\|_j}{\|\mathbf{x}\|_j} \right\} \tag{1.32}$$

which is often called the *spectral norm* of A.

Corresponding to $j = 1, 2$, and ∞ we consider the norms defined earlier in this section.

In Example 1.6 we saw that

$$\|A\|_\infty = \max_i |A_{i.}'|.$$

It can also be shown that

$$\|A\|_1 = \max_j |A_{.j}|$$

and

$$\|A\|_2 = [\lambda_{\max}(A^*A)]^{1/2}$$

where $\lambda_{\max}$ is the maximum eigenvalue of the matrix A^*A (for proof see Example 3.2).

Some properties of matrix norms are listed below

$$\begin{aligned}
&\|A\mathbf{x}\| \leqslant \|A\|\,\|\mathbf{x}\| && (\text{all } \mathbf{x}) \\
&\|A\| \geqslant 0 && (\text{with } \|A\| = 0 \text{ iff } A = 0) \\
&\|\lambda A\| = |\lambda|\,\|A\| && (\text{all scalars } \lambda) \\
&\|A + B\| \leqslant \|A\| + \|B\| && \\
&\|AB\| \leqslant \|A\|\,\|B\| && \\
&\|I\| = 1 && (I \text{ is the identity matrix}).
\end{aligned} \tag{1.34}$$

In general we shall confine ourselves to the Euclidean norm, corresponding to $i = 2$.

Lemma
Given a matrix A then

$$\|A\| \geqslant |\lambda|$$

for any eigenvalue λ of A.

Proof
Since

$$\lambda \mathbf{x} = A\mathbf{x}$$

$$|\lambda| \, \|\mathbf{x}\| = \|\lambda \mathbf{x}\| = \|A\mathbf{x}\| \leqslant \|A\| \, \|\mathbf{x}\|$$

by (1.23) and (1.34).
The result now follows.

1.11 THE TRACE FUNCTION

The *trace* (or *spur*) of a matrix $A(n \times n) = [a_{ij}]$ is the sum of the diagonal terms

$$\operatorname{tr}(A) = \sum_{i=1}^{n} a_{ii} \tag{1.35}$$

We use the notation introduced in Section 1.2. A fuller discussion is given in [B9 Section 1.3]. The (i, j)th entry of A can be written as

$$a_{ij} = \mathbf{e}_i' A \mathbf{e}_j' = \mathbf{e}_j' A' \mathbf{e}_i$$

so that

$$a_{ii} = \mathbf{e}_i' A \mathbf{e}_i$$

and

$$\operatorname{tr}(A) = \sum \mathbf{e}_i' A \mathbf{e}_i.$$

For a product we use the result

$$AB = \sum (A\mathbf{e}_j)(\mathbf{e}_j' B),$$

then

$$\begin{aligned} \operatorname{tr}(AB) &= \sum \mathbf{e}_i' AB\, \mathbf{e}_i \\ &= \sum_j \sum_i (\mathbf{e}_i' A\, \mathbf{e}_j)(\mathbf{e}_j' B\, \mathbf{e}_i) \\ &= \sum_j \sum_i a_{ij} b_{ji} \end{aligned} \tag{1.36}$$

Similarly

$$\text{tr}(BA) = \sum_i \sum_j b_{ji} a_{ij}$$

Hence

$$\text{tr}(AB) = \text{tr}(BA) \tag{1.37}$$

A useful application of (1.36) concerns complex matrices. We obtain

$$\text{tr}(AA^*) = \sum_{i,j} a_{ij} \bar{a}_{ij} \qquad (\text{since } b_{ji} = \bar{a}_{ij})$$

$$= \sum_{i,j} |a_{ij}|^2 \tag{1.38}$$

We use (1.37) to prove that similar matrices have equal traces. Consider the matrices A and $P^{-1}AP$.

Let $P^{-1}A = B$, then

$$\text{tr}(P^{-1}AP) = \text{tr}(BP) = \text{tr}(PB) \qquad (\text{by } 1.37)$$

$$= \text{tr}(PP^{-1}A) = \text{tr}\, A.$$

Hence for any non-singular matrix P,

$$\text{tr}\, A = \text{tr}(P^{-1}AP). \tag{1.39}$$

The trace of a matrix is a linear function, this is proved by showing that

$$\text{tr}(A + B) = \text{tr}\, A + \text{tr}\, B$$

and

$$\text{tr}(\alpha A) = \alpha\, \text{tr}\, A \qquad (\alpha \text{ is a scalar})$$

The trace function was introduced with reference to matrix eigenvalues in Section 1.7. We obtained an important result (1.10) restated here

$$\text{tr}\, A = b_{n-1} = \text{coef. of } \lambda^{n-1}$$

in the characteristic equation of A.

Writing the characteristic equation as in (1.11), we immediately conclude that

$$\text{tr}\, A = b_{n-1} = \lambda_1 + \lambda_2 + \ldots + \lambda_n \tag{1.42}$$

Example 1.7

Given a matrix $A = [a_{ij}]$ of order 3 having a characteristic polynomial $C(\lambda)$ show that

$$C'(\lambda) = \text{tr}\, B(\lambda)$$

where $B = [b_{ij}]$ is the adjoint of A.

Solution

$$C(\lambda) = \begin{vmatrix} \lambda - a_{11} & -a_{12} & -a_{13} \\ -a_{21} & \lambda - a_{22} & -a_{23} \\ -a_{31} & -a_{32} & \lambda - a_{33} \end{vmatrix}.$$

From Section 1.6

$$C'(\lambda) = \begin{vmatrix} 1 & -a_{12} & -a_{13} \\ 0 & \lambda - a_{22} & -a_{32} \\ 0 & -a_{23} & \lambda - a_{33} \end{vmatrix} + \begin{vmatrix} \lambda - a_{11} & 0 & -a_{13} \\ -a_{21} & 1 & -a_{23} \\ -a_{31} & 0 & \lambda - a_{33} \end{vmatrix}$$

$$+ \begin{vmatrix} \lambda - a_{11} & -a_{12} & 0 \\ -a_{21} & \lambda - a_{22} & 0 \\ -a_{31} & -a_{32} & 1 \end{vmatrix}.$$

On expanding, it follows that

$$C'(\lambda) = \begin{vmatrix} \lambda - a_{22} & -a_{32} \\ -a_{23} & \lambda - a_{33} \end{vmatrix} + \begin{vmatrix} \lambda - a_{11} & -a_{13} \\ -a_{31} & \lambda - a_{33} \end{vmatrix}$$

$$+ \begin{vmatrix} \lambda - a_{11} & -a_{12} \\ -a_{21} & \lambda - a_{12} \end{vmatrix}$$

$$= b_{11} + b_{22} + b_{33}$$

$$= \operatorname{tr} B.$$

1.12 PERMUTATION MATRICES AND IRREDUCIBLE MATRICES

1.12.1 Permutations and Permutation Matrices

The set of permuations σ_i of the set $N = \{1, 2, \ldots, n\}$ onto itself is denoted by S_n. The set S_n consists of $n!$ elements. A typical element is defined by

$$\begin{cases} \sigma(1) = i_1 \\ \sigma(2) = i_2 \\ \vdots \\ \sigma(n) = i_n \end{cases}$$

where $i_1, i_2, \ldots, i_n$ is some rearrangement of the elements of N. This permutation is often written as

$$\sigma : \begin{pmatrix} 1 & 2 & \ldots & n \\ i_1 & i_2 & \ldots & i_n \end{pmatrix}.$$

The *identity permutation* is defined as

$$\sigma : \begin{pmatrix} 1 & 2 & \ldots & n \\ 1 & 2 & \ldots & n \end{pmatrix}.$$

The *inverse permutation*, denoted by σ^{-1}, is defined by

$$\sigma^{-1}(i_r) = r \qquad (r = 1, 2, \ldots, n).$$

A *permutation matrix* of order n is defined as

$$P = P_\sigma = \begin{bmatrix} \mathbf{e}'_{i_1} \\ \mathbf{e}'_{i_2} \\ \cdot \\ \cdot \\ \cdot \\ \mathbf{e}'_{i_n} \end{bmatrix}$$

where $\mathbf{e}_1, \mathbf{e}_2, \ldots, \mathbf{e}_n$ are the unit vectors of order n (see Section 1.1). It follows that

$$P = [p_{ij}]$$

where

$$p_{ij} = \begin{cases} 1 \text{ for } j = \sigma(i) \\ 0 \text{ otherwise} \end{cases} \qquad i = 1, 2, \ldots, n$$

For example, consider

$$\sigma : \begin{pmatrix} 1 & 2 & 3 & 4 \\ 3 & 2 & 4 & 1 \end{pmatrix},$$

so that

$$\sigma(1) = 3, \quad \sigma(2) = 2, \quad \sigma(3) = 4 \text{ and } \sigma(4) = 1$$

then

$$P_\sigma = \begin{bmatrix} 0 & 0 & 1 & 0 \\ 0 & 1 & 0 & 0 \\ 0 & 0 & 0 & 1 \\ 1 & 0 & 0 & 0 \end{bmatrix}.$$

A permutation matrix is seen to have a single unit entry in each row and in each column and zeros elsewhere. As every permutation matrix P is (obviously) orthogonal, it has the property

$$P' = P^{-1}.$$

For the example above, using the definition of the inverse permutation, we have

$$\sigma^{-1} : \begin{pmatrix} 1 & 2 & 3 & 4 \\ 4 & 2 & 1 & 3 \end{pmatrix}$$

so that

$$P_{\sigma^{-1}} = \begin{bmatrix} 0 & 0 & 0 & 1 \\ 0 & 1 & 0 & 0 \\ 1 & 0 & 0 & 0 \\ 0 & 0 & 1 & 0 \end{bmatrix}.$$

We conclude that

$$P_{\sigma^{-1}} = P^{-1} = P'.$$

From the definition of P_σ, it is easily verified that

$$P_\sigma \begin{bmatrix} 1 \\ 2 \\ \vdots \\ n \end{bmatrix} = \begin{bmatrix} \sigma(1) \\ \sigma(2) \\ \vdots \\ \sigma(n) \end{bmatrix}.$$

More generally, if $\mathbf{x}$ is a vector of order n

$$P_\sigma \mathbf{x} = \mathbf{x}_\sigma$$

where $\mathbf{x}_\sigma$ is $\mathbf{x}$ whose entries (or rows) are permuted by σ.

It follows that if $A(n \times m) = [a_{ij}]$ then

$$P_\sigma A = [a_{\sigma(i),j}]$$

so that $P_\sigma A$ is A whose rows are permuted by σ.

For example, using the above permutation

$$P_\sigma A = \begin{bmatrix} 0 & 0 & 1 & 0 \\ 0 & 1 & 0 & 0 \\ 0 & 0 & 0 & 1 \\ 1 & 0 & 0 & 0 \end{bmatrix} \begin{bmatrix} a_{11} & a_{12} & a_{13} \\ a_{21} & a_{22} & a_{23} \\ a_{31} & a_{32} & a_{33} \\ a_{41} & a_{42} & a_{43} \end{bmatrix} = \begin{bmatrix} a_{31} & a_{32} & a_{33} \\ a_{21} & a_{22} & a_{23} \\ a_{41} & a_{42} & a_{43} \\ a_{11} & a_{12} & a_{13} \end{bmatrix}$$

which illustrates that $P_\sigma A$ is the matrix A whose rows have been permuted by

$$\sigma : \begin{pmatrix} 1 & 2 & 3 & 4 \\ 3 & 2 & 4 & 1 \end{pmatrix}.$$

Similarly

$$[1, 2, \ldots, n]\ P_\sigma = [\sigma^{-1}(1), \sigma^{-1}(2), \ldots, \sigma^{-1}(n)]$$

or

$$\mathbf{x}' P_\sigma = \mathbf{x}'_{\sigma^{-1}}$$

where $\mathbf{x}'_{\sigma^{-1}}$ is $\mathbf{x}'$ whose entries (or columns) are permuted by σ^{-1}. It follows that if $A(m \times n) = [a_{ij}]$, then

$$AP_\sigma = [a_{i,\sigma^{-1}(j)}]$$

so that AP_σ is A whose columns are permuted by σ^{-1}. For example, using the above permutation

$$AP_\sigma = \begin{bmatrix} a_{11} & a_{12} & a_{13} & a_{14} \\ a_{21} & a_{22} & a_{23} & a_{24} \\ a_{31} & a_{32} & a_{33} & a_{34} \end{bmatrix} \begin{bmatrix} 0 & 0 & 1 & 0 \\ 0 & 1 & 0 & 0 \\ 0 & 0 & 0 & 1 \\ 1 & 0 & 0 & 0 \end{bmatrix}$$

$$= \begin{bmatrix} a_{14} & a_{12} & a_{11} & a_{13} \\ a_{24} & a_{22} & a_{21} & a_{23} \\ a_{34} & a_{32} & a_{31} & a_{33} \end{bmatrix}$$

so that AP_σ is the matrix A whose columns have been permuted by

$$\sigma^{-1} : \begin{pmatrix} 1 & 2 & 3 & 4 \\ 4 & 2 & 1 & 3 \end{pmatrix}.$$

Finally if $A(n \times n) = [a_{ij}]$, and writing P_σ as P and $P_{\sigma^{-1}}$ as P', then

$$PAP' = [a_{\sigma(i),\sigma(j)}],$$

so that PAP' is the matrix A whose rows *and* columns are permuted by σ. This is achieved by a permutation of the *indices* $\{1, 2, \ldots, n\}$ corresponding to the n rows and n columns of the matrix $A(n \times n)$.

For example, let

$$\sigma = \begin{pmatrix} 1 & 2 & 3 \\ 2 & 3 & 1 \end{pmatrix},$$

then

$$PAP' = \begin{bmatrix} 0 & 1 & 0 \\ 0 & 0 & 1 \\ 1 & 0 & 0 \end{bmatrix} \begin{bmatrix} a_{11} & a_{12} & a_{13} \\ a_{21} & a_{22} & a_{23} \\ a_{31} & a_{32} & a_{33} \end{bmatrix} \begin{bmatrix} 0 & 0 & 1 \\ 1 & 0 & 0 \\ 0 & 1 & 0 \end{bmatrix}$$

$$= \begin{bmatrix} a_{22} & a_{23} & a_{21} \\ a_{32} & a_{33} & a_{31} \\ a_{12} & a_{13} & a_{11} \end{bmatrix}.$$

Some important concepts are now introduced.

Let $\pi, \sigma \in S_n$, then the *product* of permutations

$$\pi \circ \sigma$$

is the permutation obtained by first applying σ, then π, so that

$$(\pi \circ \sigma)(x) = \pi\{\sigma(x)\}.$$

Example 1.8
Let

$$\sigma : \begin{pmatrix} 1 & 2 & 3 \\ 3 & 1 & 2 \end{pmatrix} \quad \text{and} \quad \pi : \begin{pmatrix} 1 & 2 & 3 \\ 3 & 2 & 1 \end{pmatrix}$$

then

$$\pi \circ \sigma : \begin{pmatrix} 1 & 2 & 3 \\ 1 & 3 & 2 \end{pmatrix} \text{and} \quad \sigma \circ \pi : \begin{pmatrix} 1 & 2 & 3 \\ 2 & 1 & 3 \end{pmatrix}.$$

Another important concept is that of a *cycle.*

Definition
Let $\{i_1, i_2, \ldots, i_r\}$ be a subset of N and $\sigma \in S_n$ is such that

$$\sigma(i_1) = i_2, \sigma(i_2) = i_3, \ldots, \pi(i_{r-1}) = i_r \quad \text{and} \quad \sigma(i_r) = \mathrm{i}_1$$

and

$$\sigma(x) = x \text{ for } x \notin \{i_1, i_2, \ldots, i_r\}$$

then σ is said to have a *cycle of length r* or an *r-cycle*, denoted by $(i_1, i_2, \ldots, i_r)$.

For example

$$\begin{pmatrix} 1 & 2 & 3 & 4 & 5 & 6 \\ 1 & 5 & 3 & 2 & 4 & 6 \end{pmatrix} = (2, 5, 4) \text{ is s 3-cycle in } S_6$$

which is a shorthand notation for

$$\begin{pmatrix} 2 & 5 & 4 \\ 5 & 4 & 2 \end{pmatrix} \circ \begin{pmatrix} 1 & 2 & 3 & 4 & 5 & 6 \\ 1 & 2 & 3 & 4 & 5 & 6 \end{pmatrix}.$$

In particular if in the above definition $r = n$, σ is said to be a *full-cycle* permutation.

A generalisation of the above concepts follows:

Definition
Let $\sigma \in S_n$ and $i \in N$, then the *orbit of i under σ* are the distinct elements i, $\sigma(i)$, $\sigma^2(i) \ldots$.

The above definition makes it possible to split a permutation into orbits each of which in turn give rise to a cycle so that a permutation which is not a cycle can be split into a product of cycles.

Example 1.9
For

$$\sigma: \begin{pmatrix} 1 & 2 & 3 & 4 & 5 & 6 & 7 & 8 \\ 5 & 4 & 1 & 2 & 8 & 6 & 7 & 3 \end{pmatrix}$$

$$\sigma(1) = 5,\ \sigma^2(1) = \sigma(5) = 8,\ \sigma^3(1) = \sigma^2(5) = \sigma(8) = 3$$

and

$$\sigma^4(1) = 1.$$

The orbit of $i = 1$ is $\{1, 5, 8, 3\}$.
The orbit of 2 and 4 is $\{2, 4\}$ which gives rise to the 2-cycle $(2, 4)$.
σ leaves 6 and 7 fixed, implying that $\{6\}$ and $\{7\}$ give rise to the 1-cycles (6) and (7).
It can now be verified that

$$\sigma : (1, 5, 8, 3) \circ (2, 4) \circ (6) \circ (7)$$

which are disjoint cycles.

Often a notation is used in which the 1-cycles are omitted and the above is written as

$$\sigma : (1, 5, 8, 3) \circ (2, 4).$$

It is not difficult to prove that every permutation can be written as a product of disjoint cycles. The product is unique up to the arrangement of the factors.

If there are m cycles involved of lengths $p_1, p_2, \ldots, p_m$ respectively, then σ is said to have a *cycle structure* $[p_1, p_2, \ldots, p_m]$.

Example 1.10
Consider σ:

$$\begin{pmatrix} 1 & 2 & 3 & 4 & 5 \\ 2 & 3 & 1 & 5 & 4 \end{pmatrix}$$

corresponding to

$$p = \begin{bmatrix} 0 & 1 & 0 & 0 & 0 \\ 0 & 0 & 1 & 0 & 0 \\ 1 & 0 & 0 & 0 & 0 \\ 0 & 0 & 0 & 0 & 1 \\ 0 & 0 & 0 & 1 & 0 \end{bmatrix}.$$

σ can be written as the product of cycles

$$(2\ \ 3\ \ 1) \circ (5\ \ 4)$$

so that the cycle structure is $[3, 2]$.

Example 1.11
Consider σ:

$$\begin{pmatrix} 1 & 2 & 3 & 4 & 5 & 6 \\ 5 & 1 & 6 & 4 & 2 & 3 \end{pmatrix}.$$

σ can be written as a product of three cycles

$$(1, 5, 2) \circ (4) \circ (3, 6).$$

The cycle structure is [3, 1, 2].

We next consider an important example involving a permutation σ which is a full-cycle.

Example 1.12

Consider σ:

$$\begin{pmatrix} 1 & 2 & 3 & 4 \\ 4 & 1 & 2 & 3 \end{pmatrix}$$

corresponding to

$$P = \begin{bmatrix} 0 & 0 & 0 & 1 \\ 1 & 0 & 0 & 0 \\ 0 & 1 & 0 & 0 \\ 0 & 0 & 1 & 0 \end{bmatrix}.$$

For this matrix

$$P^4 = I.$$

As we will be interested in permutation matrices of the above structure but of various orders, we consider P of order $(n \times n)$.

For this matrix

$$P^n = I.$$

Further both the minimal and the characteristic polynomials for P are

$$\lambda^n - 1 = 0.$$

So there are n distinct eigenvalues of P, they are the nth roots of unity

$$\lambda_k = \exp i \left\{ \frac{2\pi k}{n} \right\} \qquad (k = 0, 1, \ldots, n-1).$$

We can also evaluate the corresponding eigenvectors.

Consider a matrix of form P above but of order $(n \times n)$. Notice that if $\mathbf{e}_1, \mathbf{e}_2, \ldots, \mathbf{e}_n$ are the unit vectors of order n, then for the matrix P

$$P\mathbf{e}_1 = \mathbf{e}_2, P\mathbf{e}_2 = \mathbf{e}_3, \ldots, P\mathbf{e}_{n-1} = \mathbf{e}_n, P\mathbf{e}_n = \mathbf{e}_1.$$

Let

$$\lambda = \exp i \left\{ \frac{2\pi}{n} \right\},$$

then

$$\lambda_k = \lambda^k \ (k = 0, 1, \ldots, n-1)$$

Let

$$\mathbf{x}_j = \sum_{k=1}^{n} \lambda^{kj}\, \mathbf{e}_k \qquad (j \text{ is some integer in the set } \{1, 2, \ldots, n\})$$

then

$$\begin{aligned} P\mathbf{x}_j &= \sum_{k=1}^{n} \lambda^{kj} P\mathbf{e}_k \\ &= \lambda^j \mathbf{e}_2 + \lambda^{2j}\mathbf{e}_3 + \ldots + \lambda^{(n-1)j}\, \mathbf{e}_n + \lambda^{nj}\mathbf{e}_1. \\ &= \lambda^{-j}\,[\lambda^{(n+1)j}\mathbf{e}_1 + \lambda^{2j}\mathbf{e}_2 + \lambda^{3j}\mathbf{e}_3 + \ldots + \lambda^{nj}\mathbf{e}_n] \\ &= \lambda^{n-j} \sum_{k=1}^{n} \lambda^{kj}\mathbf{e}_k \qquad\qquad (\text{since } \lambda^{-j} = \lambda^{n-j}) \\ &= \lambda^{n-j}\mathbf{x}_j. \end{aligned}$$

Hence corresponding to the eigenvalue

λ^{n-j} of P

the eigenvector is

$$\mathbf{x}_j = [\lambda^j, \lambda^{2j}, \ldots, \lambda^{nj}]' \qquad (j = 1, 2, \ldots, n)$$

Example 1.13
Find the eigenvalues and the eigenvectors of

$$P = \begin{bmatrix} 0 & 0 & 1 \\ 1 & 0 & 0 \\ 0 & 1 & 0 \end{bmatrix}.$$

Solution
In this case

$$\lambda = \exp i\left\{\frac{2\pi}{3}\right\}, \qquad \exp i\left\{\frac{4\pi}{3}\right\} \quad \text{and } 1.$$

The corresponding eigenvectors are:

$$\begin{bmatrix} \lambda^2 \\ \lambda^4 \\ \lambda^6 \end{bmatrix} = \begin{bmatrix} \lambda^2 \\ \lambda \\ 1 \end{bmatrix}, \quad \begin{bmatrix} \lambda \\ \lambda^2 \\ 1 \end{bmatrix}, \quad \text{and} \quad \begin{bmatrix} 1 \\ 1 \\ 1 \end{bmatrix}.$$

where $\lambda = -0.5 + 0.866i$ and $\lambda^2 = -0.5 - 0.866i$. ∎

It is also clear that any permutation matrix P corresponding to a full cycle permutation σ can be transformed into the matrix form described in Example 1.12 by

$$RPR'$$

where R is a permutation matrix.

For example

$$P = \begin{bmatrix} 0 & 1 & 0 & 0 \\ 0 & 0 & 1 & 0 \\ 0 & 0 & 0 & 1 \\ 1 & 0 & 0 & 0 \end{bmatrix}$$

is associated with the 4-cycle permutation

$$(2 \quad 1 \quad 4 \quad 3).$$

The matrix

$$R = \begin{bmatrix} 0 & 1 & 0 & 0 \\ 1 & 0 & 0 & 0 \\ 0 & 0 & 1 & 0 \\ 0 & 0 & 0 & 1 \end{bmatrix}$$

is such that

$$RPR'$$

is in the desired form,

We can now generalise the method used to find the eigenvalues of permutation matrix in Example 1.12 to find the eigenvalues of other permutation matrices.

We make use of two well known results

(1) PAP' and A have the same eigenvalues.
(2) The eigenvalues of a matrix A in a block diagonal form is the sum of the eigenvalues of each block; so that if for example

$$A = \left[\begin{array}{c:c:c} A_1 & 0 & 0 \\ \hdashline 0 & A_2 & 0 \\ \hdashline 0 & 0 & A_3 \end{array}\right]$$

where A_1, A_2 and A_3 are square matrices, then eigenvalues of A $= \Sigma_i$ eigenvalues A_i.

So for the above illustrative example, since the eigenvalues of

$$P \text{ and } RPR'$$

are the same, the eigenvalues of P are the ones found in Example 1.12.

Example 1.14

Find the eigenvalues of the permutation matrices considered in Examples 1.10 and 1.11.

Solution

The matrix in Example 1.10 is already in a block diagonal form

$$P = \begin{bmatrix} A_1 & 0 \\ 0 & A_2 \end{bmatrix}$$

where

$$A_1 = \begin{bmatrix} 0 & 1 & 0 \\ 0 & 0 & 1 \\ 1 & 0 & 0 \end{bmatrix} \quad \text{and} \quad A_2 = \begin{bmatrix} 0 & 1 \\ 1 & 0 \end{bmatrix}.$$

Although A_1 is not in the form of the matrix in Example 1.13, we know that the eigenvalues must be the same.

Formally, there is no difficulty in constructing the permutation matrix

$$R = \left[\begin{array}{ccc:cc} 0 & 1 & 0 & 0 & 0 \\ 1 & 0 & 0 & 0 & 0 \\ 0 & 0 & 1 & 0 & 0 \\ \hdashline 0 & 0 & 0 & 1 & 0 \\ 0 & 0 & 0 & 0 & 1 \end{array}\right]$$

such that

$$RPR' = \begin{bmatrix} A_3 & 0 \\ 0 & A_4 \end{bmatrix}$$

where A_3 is now in the desired form, whereas $A_4 = A_2$ is also in the desired form.

The matrix P corresponding to the permutation σ in Example 1.11 is not in a block diagonal form. But a rearrangement of rows and columns in such a way that the integers forming the disjoint cycles are brought together in groups will transform this matrix into such a form.

A matrix R corresponding to the permutation

$$\mu : \begin{pmatrix} 1 & 2 & 3 & 4 & 5 & 6 \\ 1 & 5 & 2 & 4 & 3 & 6 \end{pmatrix}$$

will obviously be such that

$$RPR'$$

corresponds to such a rearrangement.

Indeed for

$$P = \begin{bmatrix} 0 & 0 & 0 & 0 & 1 & 0 \\ 1 & 0 & 0 & 0 & 0 & 0 \\ 0 & 0 & 0 & 0 & 0 & 1 \\ 0 & 0 & 0 & 1 & 0 & 0 \\ 0 & 1 & 0 & 0 & 0 & 0 \\ 0 & 0 & 1 & 0 & 0 & 0 \end{bmatrix} \quad \text{and} \quad R = \begin{bmatrix} 1 & 0 & 0 & 0 & 0 & 0 \\ 0 & 0 & 0 & 0 & 1 & 0 \\ 0 & 1 & 0 & 0 & 0 & 0 \\ 0 & 0 & 0 & 1 & 0 & 0 \\ 0 & 0 & 1 & 0 & 0 & 0 \\ 0 & 0 & 0 & 0 & 0 & 1 \end{bmatrix}$$

$$RPR' = \left[\begin{array}{ccc:c:cc} 0 & 1 & 0 & 0 & 0 & 0 \\ 0 & 0 & 1 & 0 & 0 & 0 \\ 1 & 0 & 0 & 0 & 0 & 0 \\ \hdashline 0 & 0 & 0 & 1 & 0 & 0 \\ \hdashline 0 & 0 & 0 & 0 & 0 & 1 \\ 0 & 0 & 0 & 0 & 1 & 0 \end{array}\right].$$

The eigenvalues are now easily determined.

1.12.2 Irreducible Matrices

A matrix A is said to be *reducible* (or *decomposable*) if there exists a permutation matrix P such that

$$PAP' = \left[\begin{array}{c:c} A_{11} & A_{12} \\ \hdashline 0 & A_{22} \end{array}\right] \tag{1.43}$$

where A_{11} and A_{22} are *square* matrices not necessarily of the same order. If no such matrix P exists we say that A is *irreducible* (or *indecomposable*). In particular, every matrix of order (1×1) is irreducible. For example, when

$$A = \begin{bmatrix} 2 & 2 & 4 \\ 0 & 1 & 0 \\ 3 & 2 & 1 \end{bmatrix} \quad \text{and} \quad P = \begin{bmatrix} 0 & 0 & 1 \\ 1 & 0 & 0 \\ 0 & 1 & 0 \end{bmatrix},$$

then

$$PAP' = \left[\begin{array}{cc:c} 1 & 3 & 2 \\ 4 & 2 & 2 \\ \hdashline 0 & 0 & 1 \end{array}\right]$$

so that A is reducible.

Note. Some authors define A to be reducible if there exists a permucation matrix P such that

$$PAP' = \left[\begin{array}{c:c} A_{11} & 0 \\ \hdashline 0 & A_{22} \end{array}\right]$$

where A_{11} and A_{22} are square.

Most authors seem to agree on the definition of reducibility, although some write it in the equivalent form.

$$PAP' = \left[\begin{array}{c:c} A_{11} & 0 \\ \hdashline A_{21} & A_{22} \end{array}\right]$$

In fact it is sometimes useful (see [B14] p. 123) to distinguish between partly decomposable and fully decomposable matrices.

A matrix $A(n \times n)$ is said to be *partly decomposable* if it contains a zero submatrix of order $rx(n-r)$. Otherwise A is *fully indecomposable.* Of course a decomposable matrix is certainly partly decomposable, but the converse is not necessarily true.

For example

$$\begin{bmatrix} 1 & 1 \\ 1 & 0 \end{bmatrix}$$

is partly decomposable but it is not decomposable. ■

An example of an irreducible matrix is

$$A = \begin{bmatrix} 0 & 1 & 0 & 0 \\ 0 & 0 & 1 & 0 \\ 0 & 0 & 0 & 2 \\ 2 & 0 & 0 & 0 \end{bmatrix}$$

Notice that for a matrix A with all zero elements in any row, we can always find a permutation matrix P such that

$$PAP' = \left[\begin{array}{c:c} A_{11} & A_{12} \\ \hdashline 0 \ldots 0 & 0 \end{array}\right]$$

where A_{12} is a one-column matrix. Hence A is *reducible*.

Similarly a matrix A with a column of zeros is also reducible.

On the other hand, a matrix A whose elements are all positive is irreducible.

Using the rules of multiplication for partitioned matrices, it is simple to verify that A^k $(k > 1)$ is a reducible matrix whenever A is reducible.

On the other hand if A is irreducible it *does not follow* that A^k $(k > 1)$ is irreducible.

For example

$$A = \begin{bmatrix} 0 & 1 & 1 \\ 1 & 0 & 0 \\ 1 & 0 & 0 \end{bmatrix}$$

is irreducible, but

$$A^2 = \begin{bmatrix} 1 & 0 & 0 \\ 0 & 1 & 1 \\ 0 & 1 & 1 \end{bmatrix}$$

is reducible.

From the definition (1.43) we can construct a test to determine the reducibility status of a matrix A. For simplicity we shall examine in some detail the case of a matrix of order (3×3). The conclusions are valid for a matrix of order $(n \times n)$.

We use the notation

$$\langle \mathbf{x}_1, \mathbf{x}_2, \ldots, \mathbf{x}_k \rangle$$

to denote the space spanned by the vectors $\mathbf{x}_1, \mathbf{x}_2, \ldots, \mathbf{x}_k$.

Given $\mathbf{e}_j$, the jth unit vector (see Section 1.2)

$$A\mathbf{e}_j$$

is the jth column of the matrix A. If $A(3 \times 3)$ is reducible, it has all nonnegative entries and an appropriate permutation will transform it into the form (1.43), which takes one of the two forms

$$\left[\begin{array}{c:cc} X & X & X \\ \hdashline 0 & X & X \\ 0 & X & X \end{array}\right] \quad \text{or} \quad \left[\begin{array}{cc:c} X & X & X \\ X & X & X \\ \hdashline 0 & 0 & X \end{array}\right]$$

where X is a nonnegative number.

In the first case

$$\langle A\mathbf{e}_1 \rangle \subset \langle \mathbf{e}_1 \rangle \quad \text{becomes} \quad \langle A\mathbf{e}_1 \rangle = \langle \mathbf{e}_1 \rangle.$$

In the second case

$$\langle A\mathbf{e}_1, A\mathbf{e}_2 \rangle \subset \langle \mathbf{e}_1, \mathbf{e}_2 \rangle \quad \text{becomes} \quad \langle A\mathbf{e}_1, A\mathbf{e}_2 \rangle = \langle \mathbf{e}_1, \mathbf{e}_2 \rangle.$$

Notice that because the matrix considered is assumed to have a positive element in each row and each column (otherwise there is no need for a test), we always have

$$\langle A\mathbf{e}_1, A\mathbf{e}_2, A\mathbf{e}_3\rangle = \langle \mathbf{e}_1, \mathbf{e}_2, \mathbf{e}_3\rangle.$$

So in general we conclude that A is reducible if there exists a proper subset $\{1, 2, \ldots, r\}$ of $\{1, 2, \ldots, n\}$ such that

$$\langle A\mathbf{e}_1, A\mathbf{e}_2, \ldots, A\mathbf{e}_r\rangle \subset \langle \mathbf{e}_1, \mathbf{e}_2, \ldots, \mathbf{e}_r\rangle$$

Of course if A is in the form (1.43), there is no need to test for reducibility. But we have seen above that a simultaneous rearrangement of the rows and columns of a matrix A is equivalent to a permutation of the matrix indices $\{1, 2, \ldots, n\}$. Hence in the general case we conclude that A is reducible if there exists a proper subset $\{k_1, k_2, \ldots, k_r\}$ of $\{1, 2, \ldots, n\}$ such that

$$\langle A\mathbf{e}_{k_1}, A\mathbf{e}_{k_2}, \ldots, A\mathbf{e}_{k_r}\rangle \subset \langle \mathbf{e}_{k_1}, \mathbf{e}_{k_2}, \ldots, \mathbf{e}_{k_r}\rangle \tag{1.44}$$

For example, consider

$$A_1 = \begin{bmatrix} 1 & 1 & 1 \\ 0 & 1 & 0 \\ 1 & 1 & 1 \end{bmatrix} \quad \text{and} \quad A_2 = \begin{bmatrix} 0 & 1 & 0 \\ 0 & 0 & 1 \\ 1 & 0 & 0 \end{bmatrix}.$$

Since $\langle A_1\mathbf{e}_1, A_1\mathbf{e}_3\rangle = \langle \mathbf{e}_1, \mathbf{e}_3\rangle$, A_1 is reducible.

Since $A_2\mathbf{e}_j = \mathbf{e}_{j-1}$ ($j = 1, 2, 3$ with $\mathbf{e}_0$ to be interpreted as $\mathbf{e}_3$) it is impossible to find a proper subset of $\{1, 2, 3\}$ to satisfy (1.44) so that A_2 is irreducible.

In Section 1.13 we discuss a simple graphical method for determining the reducibility status of a matrix.

1.13 SOME ASPECTS OF THE THEORY OF GRAPHS

If $A = [a_{ij}]$ is a nonnegative matrix, the pattern of the zero and non-zero entries a_{ij} in A determines the pattern of the zero and non-zero entries $a_{ij}^{(k)}$ in A^k, the kth power of A.

When studying the eigenvalues and eigenvectors of nonnegative matrices, these patterns are of great importance. In this context, the concept of the directed graph associated with a matrix A is very useful.

Definitions and Notation

Give a matrix $A = [a_{ij}]$ of order $(n \times n)$, and the index set $N = \{1, 2, \ldots, n\}$ there corresponds a set of points (called *nodes* or *vertices*)

$$V = \{P_1, P_2, \ldots, P_n\}$$

and a set of lines joining pairs of these points.

A *directed line* from the point P_i to the point P_j will be denoted by $\overrightarrow{P_iP_j}$ or by $\overrightarrow{(i, j)}$.

A *directed graph* or *digraph* consists of the set V of points together with a subset of the set of directed lines joining pairs of these points.

Fig. 2 shows typical examples of digraphs. Notice that (b) involves both $\overrightarrow{(2,3)}$ and $\overrightarrow{(3,2)}$.

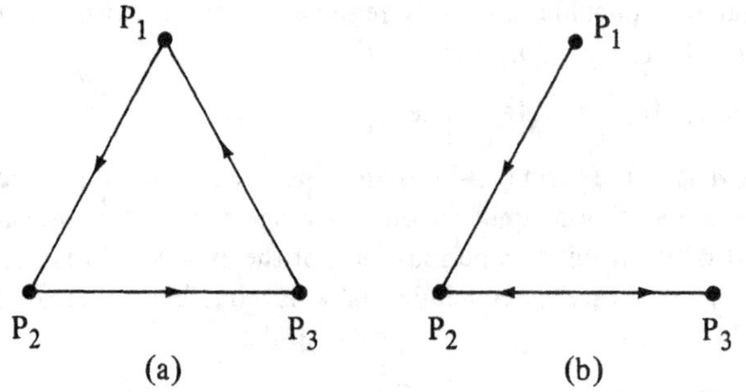

Fig. 2

A *directed path* from P_i and P_j is a collection of lines

$$\overrightarrow{P_iP_s}, \overrightarrow{P_sP_t}, \ldots, \overrightarrow{P_kP_j}$$

where all the points $P_i, P_s, P_t, \ldots, P_k, P_j$ are different from each other. If not all the points are distinct, then the collection of lines is called a *redundant chain* from P_i to P_j.

A *cycle* of a graph is a path from P_i to P_i and the *length* of a path is the number of lines in it. In particular a cycle of length 1 is called a *loop*.

A digraph is said to be *strongly connected* or *strong* if for every pair of points P_i and P_j $(i, j \in N)$, there exists a directed path from P_i to P_j *and* a directed path from P_j to P_i.

In particular, a graph with one point and no lines is strongly connected.

We can associate a digraph with a nonnegative matrix $A = [a_{ij}]$ in the following way:

For every $a_{ij} > 0$ we connect the nodes P_i and P_j by the directed line

$$\overrightarrow{P_iP_j}.$$

Having carried out the above operation for all $i, j \in N$, the resulting digraph is denoted by $G(A)$.

For example, given

$$A = \begin{bmatrix} 1 & 2 & 0 & 2 \\ 1 & 0 & 3 & 2 \\ 1 & 0 & 1 & 1 \\ 2 & 2 & 1 & 2 \end{bmatrix}$$

$G(A)$ is shown in Fig. 3.

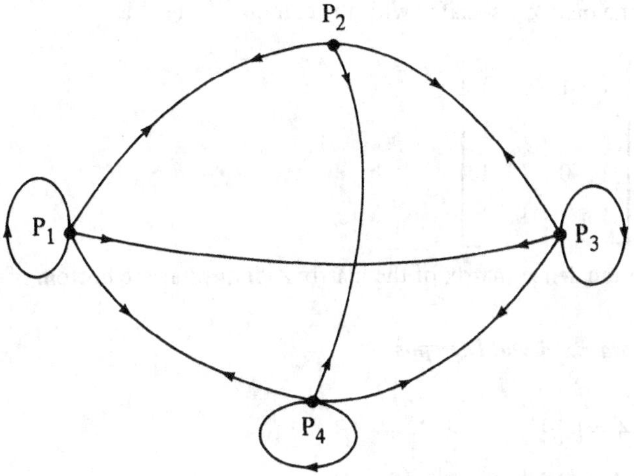

Fig. 3

There are various ways of associating a matrix with a graph. We use the following definition:

$$a_{ij} = 1 \text{ if } \overrightarrow{P_iP_j} \in G(A)$$

$$a_{ij} = 0 \text{ otherwise.}$$

A is a binary matrix (having entries 0 and 1 only). As already pointed out, it is the pattern of the zero and non-zero elements of the powers of a matrix which is of great interest to us, so a binary matrix is meaningful for our purposes. In fact, there are many practical applications of such matrices, for further details see [10].

Since it is the pattern which is of interest we can use Boolean operations, implying the 'ordinary' matrix multiplication together with 'addition' $\oplus$ defined by

$$1 \oplus 1 = 1$$

Note. The Boolean addition must not be confused with the 'direct sum' of matrices

$$A_1 \oplus A_2$$

for which the same notation is used.

We frequently replace a nonnegative matrix $B = [b_{ij}]$ by a binary matrix $A = [a_{ij}]$ called the *incidence matrix of* B, by defining

$$a_{ij} = 1 \text{ if } b_{ij} > 0$$

and

$$a_{ij} = 0 \text{ if } b_{ij} = 0.$$

Notice that the matrix associated with the digraph of Fig. 3 is

$$\begin{bmatrix} 1 & 1 & 0 & 1 \\ 1 & 0 & 1 & 1 \\ 1 & 0 & 1 & 1 \\ 1 & 1 & 1 & 1 \end{bmatrix}$$

which is the incidence matrix of the matrix A defined at the bottom of page 56.

Powers of a Matrix A and Digraphs

Let

$$A = [a_{ij}]$$

be a nonnegative matrix of order $(n \times n)$.

By definition, the non-zero entires of A indicate all the paths of length 1 in $G(A)$. Considering a loop to be a redundant chain of length 1, A gives all the paths and redundant chains of length 1 in $G(A)$.

Next consider

$$A^2 = [a_{ij}^{(2)}].$$

We use Boolean operations.

Since

$$a_{ij}^{(2)} = \sum_k a_{ij} a_{kj}$$

then $a_{ij}^{(2)} = 1$ whenever there is at least one point P_k such that the directed lines $\overrightarrow{(i, k)}$ and $\overrightarrow{(k, j)}$ both belong to $G(A)$.

So $a_{ij}^{(2)} = 1$ whenever there is a path from P_i to P_j of length 2 in $G(A)$. Also $a_{ij}^{(2)} = 1$ if there is a redundant chain of length 2 from P_i to P_j. For example the term $a_{22}a_{23}$ involves the path $\overrightarrow{P_2P_2}$, $\overrightarrow{P_2P_3}$ · $\overrightarrow{P_2P_2}$ implies traversing the loop about the node P_2. By a generalisation of the above argument it follows that

$$a_{ij}^{(k)} = 1$$

whenever there is a directed path or a redundant chain of length k from i to j.

For example given

$$A = \begin{bmatrix} 0 & 0 & 0 & 1 \\ 1 & 0 & 0 & 0 \\ 0 & 1 & 0 & 0 \\ 0 & 0 & 1 & 0 \end{bmatrix}$$

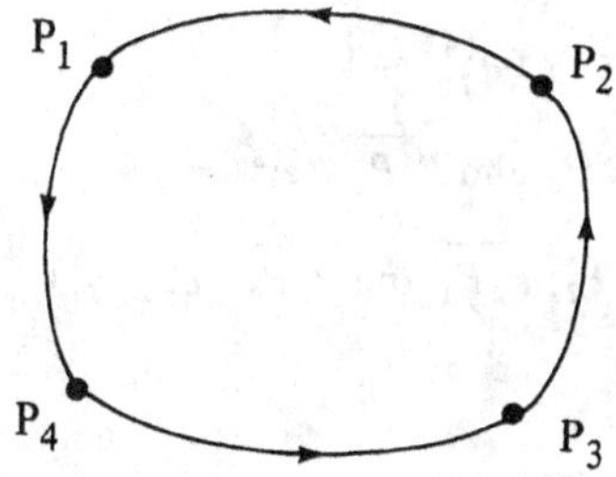

Fig. 4

$G(A)$ is as shown in Fig. 4. To obtain A^3 we note that from P_1 a path of length 3 ends at P_2, hence $a_{12}^{(3)} = 1$ and $a_{1j}^{(3)} = 0\ (j = 1, 3, 4)$. Similarly $a_{23}^{(3)} = 1$, $a_{34}^{(3)} = 1$ and $a_{41}^{(3)} = 1$. Hence

$$A^3 = \begin{bmatrix} 0 & 1 & 0 & 0 \\ 0 & 0 & 1 & 0 \\ 0 & 0 & 0 & 1 \\ 1 & 0 & 0 & 0 \end{bmatrix}.$$

Another example we consider is

$$A = \begin{bmatrix} 0 & 0 & 0 & 1 \\ 1 & 1 & 0 & 0 \\ 0 & 1 & 0 & 0 \\ 0 & 0 & 1 & 0 \end{bmatrix}.$$

$G(A)$ is as shown in Fig. 5. To obtain A^3 we consider (as an illustration) all paths of length 3 from P_2. There are 3 redundant chains:

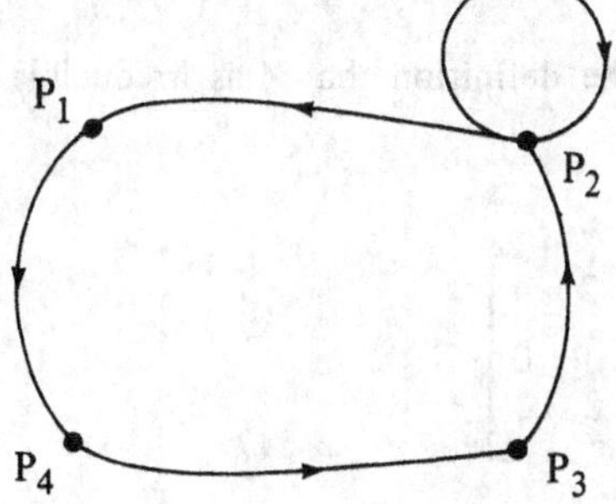

Fig. 5

3 times round the loop at P_2 : $a_{22}^{(3)} = 1$

2 times round the loop at P_2 and $\overrightarrow{P_2P_1}$: $a_{21}^{(3)} = 1$

1 time round the loop at P_2, $\overrightarrow{P_2P_1}$ and $\overrightarrow{P_2P_4}$: $a_{24}^{(3)} = 1$

also the path:

$$\overrightarrow{P_2P_1}, \overrightarrow{P_1P_4}, \overrightarrow{P_4P_3} : a_{23}^{(3)} = 1.$$

Similar considerations for the other nodes leads us to conclude that

$$A^3 = \begin{bmatrix} 0 & 1 & 0 & 0 \\ 1 & 1 & 1 & 1 \\ 1 & 1 & 0 & 1 \\ 1 & 1 & 0 & 0 \end{bmatrix}.$$

The digraph and reducibility of a matrix A

It will be shown in Section 4.5 that a nonnegative matrix $A = [a_{ij}]$ is *irreducible* if and only if for each (i, j) there exists an integer $k > 0$ such that

$$a_{ij}^{(k)} > 0.$$

Notice that k is a function of (i, j).

If $k(i, j) = C$ (a constant) for all (i, j), this means that

$$A^{(k)} > 0$$

and A is then called *primitive* (see Section 4.4.2).

But $a_{ij}^{(k)} > 0$ if and only if there exists a line sequence from P_i to P_j for each (i, j).

It follows from the definition that A is irreducible if and only if $G(A)$ is strongly connected.

For example for

$$A = \begin{bmatrix} 2 & 2 & 4 \\ 0 & 1 & 0 \\ 3 & 2 & 1 \end{bmatrix}$$

$G(A)$ is as shown in Fig. 6. Since there is no path from P_2 to P_1 or from P_2 to P_3 the matrix is reducible.

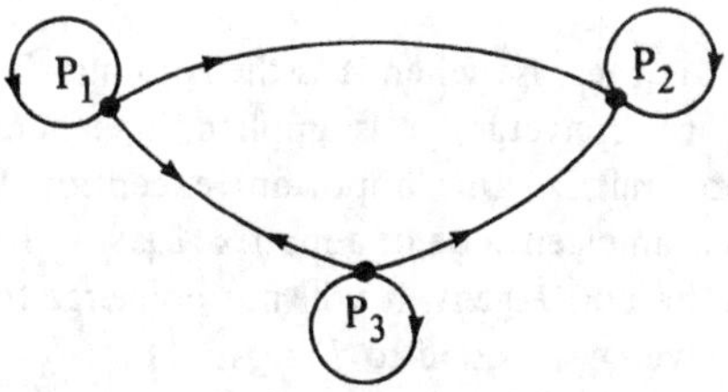

Fig. 6

On the other hand, when

$$A = \begin{bmatrix} 0 & 1 & 0 & 0 \\ 0 & 0 & 1 & 0 \\ 0 & 0 & 0 & 2 \\ 2 & 0 & 0 & 0 \end{bmatrix},$$

$G(A)$ is as shown in Fig. 7. $G(A)$ is strongly connected, hence A is irreducible.

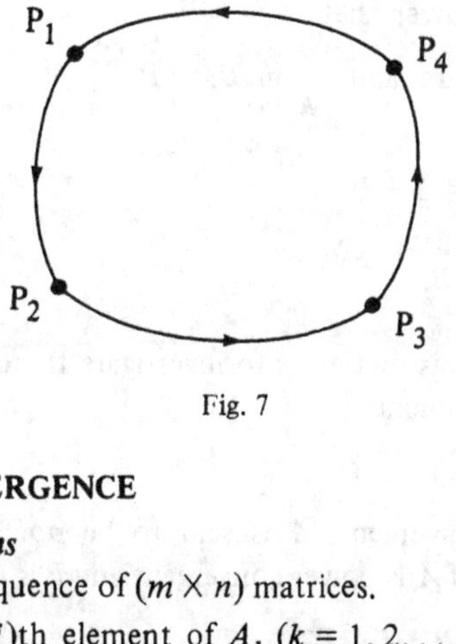

Fig. 7

1.14 MATRIX CONVERGENCE

Notation and Definitions

Let $A_1, A_2, \ldots$ be a sequence of $(m \times n)$ matrices.

Let $a_{ij}^{(k)}$ be the (i, j)th element of A_k $(k = 1, 2, \ldots)$. (Note the change in notation; up till now we have used $a_{ij}^{(k)}$ to denote the (i, j)th element in A^k.)

$A = [a_{ij}]$ is said to be the *limit* of the sequence if

$$a_{ij}^{(k)} \to a_{ij}$$

as $k \to \infty$ for each (i, j).

If the limit exists we say that the sequence $A_1, A_2, \ldots$ *converges* to A and write

$$A = \lim_{k \to \infty} A_k$$

where

$$A = [a_{ij}].$$

Of particular interest is the case when A is the zero matrix. So when a sequence of matrices is said to converge, it is implied in general that the sequence converges to the zero matrix. One important exception to this rule which we shall consider is when an eigenvalue of a matrix A is $\lambda = 1$. It will be shown that although A may still be convergent, it will not converge to the zero matrix. If a sequence does not converge, it is said to *diverge*.

For example

$$\lim_{n\to\infty} \begin{bmatrix} (n-1)/n & 3^n/n! \\ 2 & [1+(1/n)]^n \end{bmatrix} = \begin{bmatrix} 1 & 0 \\ 2 & e \end{bmatrix}$$

and

$$\lim_{n\to\infty} \begin{bmatrix} 2 \\ 1+1/n \\ 2-2/n \end{bmatrix} = \begin{bmatrix} 2 \\ 1 \\ 2 \end{bmatrix}.$$

Assuming that all matrices involved are of order $(m \times n)$ the following results are quite simple to prove: Given that

$$\lim_{k\to\infty} A_k = A \quad \text{and} \quad \lim_{k\to\infty} B_k = B.$$

(1) $\lim\limits_{k\to\infty} (A_k + B_k) = A + B$

(2) $\lim\limits_{k\to\infty} (A_k B_k) = AB.$

Our principal aim in this section is to investigate the convergence (to the zero matrix) of the power sequence

$$A, A^2, A^3, \ldots$$

If this sequence is convergent, A is said to be *power convergent*. It is not difficult to prove that if A is power convergent and P is non-singular, then

$$PAP^{-1}, PA^2P^{-1}, PA^3P^{-1}, \ldots$$

is also power convergent and conversely.

Finally if A is in a block diagonal form so that

$$A = A_1 \oplus A_2 \oplus \ldots \oplus A_r$$

where the $A_i (i = 1, 2, \ldots, r)$ are the blocks on the diagonal, then

(1) A is power convergent if and only if each A_i is power convergent.

(2) $\lim\limits_{k\to\infty} A^k = \lim\limits_{k\to\infty} A_1^k \oplus \lim\limits_{k\to\infty} A_2^k \oplus \ldots \oplus \lim\limits_{k\to\infty} A_r^k.$

Note: $\oplus$ is the 'direct' sum (see [B7] p. 191).

A matrix in a block diagonal form, with *Jordan blocks* $J_r(\lambda)$ along the diagonal, is called a *Jordan matrix*.

A Jordan block $J_r(\lambda)$ is a matrix of order $(r \times r)$ such that

(1) each diagonal entry is λ_1
(2) each superdiagonal entry is 1
(3) every other entry is 0.

For example

$$J_3(\lambda_1) = \begin{bmatrix} \lambda_1 & 1 & 0 \\ 0 & \lambda_1 & 1 \\ 0 & 0 & \lambda_1 \end{bmatrix}.$$

The importance of the Jordan matrix is due to the following fundamental result in matrix theory.

Every matrix $A(n \times n)$ is similar to some Jordan matrix.

So that given A, there exists a matrix P such that

$$J = PAP^{-1} = J_{r_1}(\lambda_1) \oplus J_{r_2}(\lambda_2) \oplus \ldots \oplus J_{r_s}(\lambda_s)$$

where

$$r_1 + r_2 + \ldots + r_s = n$$

and

$$\lambda_1, \lambda_2, \ldots, \lambda_s$$

are the distinct eigenvalues of A.

Because J is in a block diagonal form

$$J^k = PA^kP^{-1} = J^k_{r_1}(\lambda_1) \oplus J^k_{r_2}(\lambda_2) \oplus \ldots \oplus J^k_{r_s}(\lambda_s)$$

So to evaluate J^k we need to find $J^k_r(\lambda)$ for every Jordan block $J_r(\lambda)$.

We note that

$$J_r(\lambda) = \lambda I + J(0)$$

where I is of order $(r \times r)$.

$J(0)$ is a matrix of order $(r \times r)$ such that

(1) each superdiagonal entry is 1
(2) every other entry is 0.

For example

$$J_3(\lambda) = \begin{bmatrix} \lambda & 0 & 0 \\ 0 & \lambda & 0 \\ 0 & 0 & \lambda \end{bmatrix} + \begin{bmatrix} 0 & 1 & 0 \\ 0 & 0 & 1 \\ 0 & 0 & 0 \end{bmatrix}.$$

Notice that $J_r^k(0) = [0]$ for all $k \geqslant r$.

It follows that

$$J_r^k(\lambda) = [\lambda I + J(0)]^k$$
$$= \lambda^k I + k\lambda^{k-1}J(0) + \ldots + \binom{k}{j}\lambda^{k-j}J^j_{(0)} + \ldots + J^k_{(0)}$$
$$= \lambda^k I + k\lambda^{k-1}J(0) + \ldots + \binom{k}{r-1}\lambda^{k-r+1}J^{r-1}_{(0)} \cdot$$

So $J_r^k(\lambda)$ has the following form:

$$J_r^k(\lambda) = \begin{bmatrix} \lambda^k & \binom{k}{1}\lambda^{k-1} & \ldots & \binom{k}{k-r+1}\lambda^{k-r+1} \\ 0 & \lambda^k & \ldots & \binom{k}{k-r}\lambda^{k-r} \\ \vdots & & & \\ 0 & 0 & \ldots & \lambda^k \end{bmatrix}$$

where

$$\binom{k}{r} = \frac{k!}{r!(k-r)!}$$

Theorem

$J_r(\lambda)$ is power convergent if and only if $|\lambda| < 1$ and $r > 1$.

Proof

From the theory of Complex Analysis it is well known that for $k \geqslant r$

$$\lim_{k\to\infty} \binom{k}{r}\lambda^{k-r} = 0$$

whenever $|\lambda| < 1$.

If

$$|\lambda| = 1 \quad \text{and} \quad \lim_{k\to\infty} \lambda^k = L$$

then

$$L = \lim_{k\to\infty} \lambda^{k+1} = \lambda \lim_{k\to\infty} \lambda^k = \lambda L$$

hence $\lambda = 1$.

Given $\lambda = 1$, assume that $r > 1$, then the convergence of $J_r^k(1)$ implies that

$$\lim_{k\to\infty} k 1^{k-1}$$

exists. As this is not true, it follows that $r = 1$.

So in this case, when

$$\lambda = 1, r = 1$$

then

$$J_r(\lambda) = J_1(1) = [1]$$

So a Jordan block corresponding to $\lambda = 1$ in a converging Jordan matrix must be of order (1×1).

Assume now that $J_r(\lambda)$ is convergent. Then from the form for $J_r(\lambda)$ above

$$\lim_{k \to \infty} \lambda k$$

exists, so that $|\lambda| \leqslant 1$.

If $|\lambda| = 1$, then $r = 1$.

Conversely if $|\lambda| < 1$, then

$$\lim_{k \to \infty} \binom{k}{r} \lambda^{n-r} = 0$$

so that

$$\lim_{k \to \infty} [J_r(\lambda)]^k = 0.$$

If $|\lambda| = 1$, then $\lambda = 1 = r$.

From the above discussion we conclude that a matrix A having eigenvalues $\lambda_i (i = 1, 2, \ldots, n)$ is convergent if

(1) $|\lambda_i| < 1$ (all i)

or

(2) if $|\lambda| = 1$, then $\lambda = 1$ and the corresponding Jordan block is $[1]$ of order (1×1). (In this case $J_r^k(1)$ is convergent but *not* to the zero matrix.)

Series of Matrices

Given a sequence of matrices $\{A_k\}$, then the sequence of the partial sums

$$A_0, A_0 + A_1, A_0 + A_1 + A_3, \ldots, \sum_{r=0}^{m} A_r, \ldots$$

is denoted by

$$\sum_{r=0}^{\infty} A_r.$$

Assuming that the series converges, the above notation is also used for the limit.

We shall not discuss here the theory of matrix series convergence but we shall make use of the following result:

Theorem
If $A(n \times n)$ is a (complex) matrix such that

$$\rho(A) < 1$$

then

(1) $[I - A]$ is non-singular, and

(2) $[I - A]^{-1} = \sum_{r=0}^{\infty} A^r \quad (A^0 = I)$

Conversely, if

$$\sum_{r=0}^{\infty} A^r$$

converges, then $\rho(A) < 1$.

Proof
(1) Assuming that $\rho(A) < 1$, then

$$[I - A]$$

is non-singular.
(2) Now

$$[I - A]\,[I + A + A^2 + \ldots + A^k] = I - A^{k+1}.$$

Premultiplying by $[I - A]^{-1}$, results in

$$I + A + A^2 + \ldots + A^k = [I - A]^{-1} - [I - A]^{-1} A^{k+1}.$$

Now let $k \to \infty$, then since

$$\lim_{k \to \infty} A^{k+1} = 0,$$

$$\sum_{k=0}^{\infty} A^k = [I - A]^{-1}.$$

Conversely, assume λ is an eigenvalue of A and $\mathbf{x}$ the corresponding eigenvector, then

$$[I + A + A^2 + \ldots]\,\mathbf{x} = (1 + \lambda + \lambda^2 + \ldots)\,\mathbf{x}$$

Since the matrix sum is convergent, so must be the series on the right-hand side for any eigenvalue λ of A.

But from complex analysis we know that the condition for convergence is that

$$|\lambda| < 1.$$

From which it follows that

$$\rho(A) < 1.$$

2

Some Matrix Types

2.1 INTRODUCTION

In this chapter we explore some useful properties of various types of matrices. In a way this chapter is a continuation of the previous one, since once again 'preliminaries' to the main theme of this book are considered. Nevertheless the concepts involved are important, indeed essential in the development of the matrix theory as presented here.

At times we are concerned with the properties of matrices having complex elements. The equivalent matrices having real elements only are well covered in elementary matrix theory. As examples the real element equivalent of a *Hermitian* matrix is a symmetric matrix, whereas the real element equivalent of a *Unitary* matrix is an Orthogonal matrix.

2.2 UNITARY MATRICES

A matrix $U(n \times n)$ having complex elements which satisfies the equation

$$U^*U = I = UU^* \tag{2.1}$$

is said to be *unitary*.

It will be remembered that $U^* = \bar{U}'$ is the conjugate transpose (or *tranjugate*) of U.

For example

$$\frac{1}{5}\begin{bmatrix} -1+2i & -4-2i \\ 2-4i & -2-i \end{bmatrix}$$

is a unitary matrix.

If U has real elements only and satifies the condition (2.1), then U is an *orthogonal* matrix.

With the notation introduced in Section 1.2, the product UU^* can be written in partitioned form as

$$\begin{bmatrix} (U_{1\cdot})' \\ (U_{2\cdot})' \\ \vdots \\ (U_{n\cdot})' \end{bmatrix} [U^*_{\cdot 1}\, U^*_{\cdot 2} \, \ldots \, U^*_{\cdot n}] = I.$$

The (i, j)th element of this identity is the inner product

$$(U_{i\cdot}, U^*_{\cdot j}),$$

where $U^*_{\cdot j}$ is the jth column of U^*, that is the jth row of $\bar{U}$ which is $\bar{U}_j$. Hence

$$(U_{i\cdot}, \bar{U}_{j\cdot}) = \delta_{ij}.$$

This shows that the *rows of a unitary matrix U form a set of orthonormal vectors* (see Section 1.10).

A similar analysis of

$$U^*U = I$$

shows that the *columns of a unitary matrix form a set of orthonormal vectors*. Conversely, if the vectors of order n

$$\{\mathbf{x}_1, \mathbf{x}_2, \ldots, \mathbf{x}_n\}$$

are orthonormal, then the matrix U having these vectors as columns satisfies (2.1), hence it is unitary. An immediate consequence of (2.1) is that

$$U^{-1} = U^* \tag{2.2}$$

If U has real entries only, then (2.2) is equivalent to $U' = U^{-1}$ or $UU' = U'U = I$. Then U is called an *orthogonal* matrix.

Further important properties of unitary matrices are as follows:

$$|U| = 1 \tag{2.3}$$

UZ is unitary if U and Z are unitary (2.4)

To prove (2.3), we use (2.1) and the result from Section 1.5 that

$$|UU^*| = |U|\,|U^*|.$$

Hence

$$|I| = 1 = |U|\,|U^*| = |U|^2.$$

The result follows.

To prove (2.4) consider the product

$$(UZ)^*\,(UZ) = Z^*U^*UZ$$
$$= Z^*IZ = I \quad \text{(by 2.1)}$$

The result follows from the definition (2.1).

We next discuss the properties of a unitary matrix considered as a linear transformation.

We first consider a general result of scalar products.

By definition (1.25)

$$(A\mathbf{x}, B\mathbf{y}) = (A\mathbf{x})^*\,B\mathbf{y} = (\mathbf{x}^*A^*B)\mathbf{y} = (B^*A\mathbf{x})^*\mathbf{y}$$
$$= (B^*A\mathbf{x}, \mathbf{y}) \tag{2.5}$$

From (2.5), it follows that

$$(U\mathbf{x}, U\mathbf{y}) = (U^*U\mathbf{x}, \mathbf{y}) = (\mathbf{x}, \mathbf{y}). \tag{2.6}$$

and in particular

$$(U\mathbf{x}, U\mathbf{x}) = (\mathbf{x}, \mathbf{x}). \tag{2.7}$$

By (1.27) this implies that the length of vectors (the Euclidean norm) under a unitary transformation is invariant, that is

$$\|U\mathbf{x}\| = \|\mathbf{x}\| \tag{2.8}$$

Notice that

$$\|U\|_2 = \max_{\mathbf{x}\neq\mathbf{0}} \frac{\|U\mathbf{x}\|_2}{\|\mathbf{x}\|_2} = \frac{\|\mathbf{x}\|_2}{\|\mathbf{x}\|_2} = 1. \tag{2.9}$$

The angle θ between two vectors $\mathbf{x}$ and $\mathbf{y}$ is defined by

$$\cos\theta = \frac{(\mathbf{x}, \mathbf{y})}{\|\mathbf{x}\|\,\|\mathbf{y}\|}$$

After a transformation by U, the corresponding angle between the two vectors $U\mathbf{x}$ and $U\mathbf{y}$ is given by

$$\frac{(U\mathbf{x}, U\mathbf{y})}{\|U\mathbf{x}\|\,\|U\mathbf{y}\|} = \frac{(\mathbf{x}, \mathbf{y})}{\|\mathbf{x}\|\,\|\mathbf{y}\|} \tag{2.10}$$

(2.8) and (2.10) imply that under a unitary transformation *both lengths and angles are preserved.*

We can now deduce an important property of the eigenvalues of unitary matrices.

If λ is an eigenvalue of U, and $\mathbf{x}$ is the corresponding eigenvector, then

$$U\mathbf{x} = \lambda\mathbf{x}$$

so that

$$\|U\mathbf{x}\| = \|\lambda\mathbf{x}\|.$$

But $\|U\mathbf{x}\| = \|\mathbf{x}\|$ by (2.8) and $\|\lambda\mathbf{x}\| = |\lambda|\,\|\mathbf{x}\|$ by (1.23).

It follows that

$$|\lambda| = 1. \tag{2.11}$$

It is interesting to consider what is the type of the transformation defined by

$$T : \mathbf{y} = U\mathbf{x}$$

where U is an orthogonal matrix and $\mathbf{x}$ and $\mathbf{y}$ are position vectors in a Cartesian coordinate system.

Since T preserves both lengths and angles, it represents either

(1) a rotation about an axis through the origin
(2) a reflexion about an axis through the origin or
(3) a rotation followed by a reflexion of type (1) and (2).

Conversely, a rotation of a coordinate system is represented by a unitary (orthogonal) transformation. For example if the point $P(x_1, x_2)$ is rotated positively to a point $Q(y_1, y_2)$, through an angle θ, then

$$\begin{bmatrix} y_1 \\ y_2 \end{bmatrix} = \begin{bmatrix} \cos\theta & -\sin\theta \\ \sin\theta & \cos\theta \end{bmatrix} \begin{bmatrix} x_1 \\ x_1 \end{bmatrix}.$$

The matrix of rotation is orthogonal.

It has been pointed out that the rows and columns of a unitary matrix form sets of orthonormal vectors. It is worthwhile considering such sets of vectors in greater detail. The reader is probably familiar with the famous *Gram–Schmidt* orthogonalisation process which, for completeness, will now be discussed.

Statement of the Process

From a set of linearly independent set of vectors $\{\mathbf{y}_1, \mathbf{y}_2, \ldots, \mathbf{y}_n\}$, $\mathbf{y}_i \in V^{(n)}$ where $V^{(n)}$ is a n-dimensional space, we can construct an orthonormal set. $\{\mathbf{x}_1, \mathbf{x}_2, \ldots, \mathbf{x}_n\}$, $\mathbf{x}_i \in V^{(n)}$.

Construction of the orthonormal set of vectors

We first choose

$$\mathbf{z}_1 = \mathbf{y}_1$$

and then choose

$$\mathbf{x}_1 = \frac{\mathbf{z}_1}{\|\mathbf{z}_1\|} = \frac{\mathbf{z}_1}{(\mathbf{z}_1, \mathbf{z}_1)} .$$

Next let $\mathbf{z}_2 = \mathbf{y}_2 - \alpha_1 \mathbf{x}_1$, and choose the scalar α_1 so that $\mathbf{x}_1$ and $\mathbf{z}_2$ are orthogonal, that is

$$(\mathbf{x}_1, \mathbf{z}_2) = (\mathbf{x}_1, \mathbf{y}_2) - \alpha_1(\mathbf{x}_1, \mathbf{x}_1) = 0.$$

It follows that $\alpha_1 = (\mathbf{x}_1, \mathbf{y}_2)$.

Now let

$$\mathbf{x}_2 = \frac{\mathbf{z}_2}{\|\mathbf{z}_2\|} = \frac{\mathbf{z}_2}{(\mathbf{z}_2, \mathbf{z}_2)} .$$

Similarly, let

$$\mathbf{z}_3 = \mathbf{y}_3 - \alpha_2 \mathbf{x}_2 - \beta_1 \mathbf{x}_1$$

where α_2 and β_1 are chosen so that $\mathbf{z}_3$ is orthogonal to $\mathbf{x}_2$ and $\mathbf{x}_1$, hence

$$\begin{aligned}(\mathbf{x}_2, \mathbf{z}_3) = 0 &= (\mathbf{x}_2, \mathbf{y}_3) - \alpha_2(\mathbf{x}_2, \mathbf{x}_2) - \beta_1(\mathbf{x}_2, \mathbf{x}_1) \\ &= (\mathbf{x}_2, \mathbf{y}_3) - \alpha_2\end{aligned}$$

and

$$\begin{aligned}(\mathbf{x}_1, \mathbf{z}_3) = 0 &= (\mathbf{x}_1, \mathbf{y}_3) - \alpha_2(\mathbf{x}_1, \mathbf{x}_2) - \beta_1(\mathbf{x}_1, \mathbf{x}_1) \\ &= (\mathbf{x}_1, \mathbf{y}_3) - \beta_1 .\end{aligned}$$

Hence $\alpha_2 = (\mathbf{x}_2, \mathbf{y}_3)$ and $\beta_1 = (\mathbf{x}_1, \mathbf{y}_3)$.

Now let

$$\mathbf{x}_3 = \frac{\mathbf{z}_3}{\|\mathbf{z}_3\|} = \frac{\mathbf{z}_3}{(\mathbf{z}_3, \mathbf{z}_3)} .$$

To recapitulate, the substitution

$$\mathbf{z}_1 = \mathbf{y}_1 \quad \text{and} \quad \mathbf{x}_1 = \frac{\mathbf{z}_1}{\|\mathbf{z}_1\|}$$

$$\mathbf{z}_2 = \mathbf{y}_2 - (\mathbf{x}_1, \mathbf{y}_2)\,\mathbf{x}_1 \quad \text{and} \quad \mathbf{x}_2 = \frac{\mathbf{z}_2}{\|\mathbf{z}_2\|}$$

$$\mathbf{z}_3 = \mathbf{y}_3 - (\mathbf{x}_2, \mathbf{y}_3)\mathbf{x}_2 - (\mathbf{x}_1, \mathbf{y}_3)\mathbf{x}_1 \quad \text{and} \quad \mathbf{x}_3 = \frac{\mathbf{z}_3}{\|\mathbf{z}_3\|}$$

and generally,

$$\mathbf{z}_j = \mathbf{y}_j - (\mathbf{x}_{j-1}, \mathbf{y}_j)\mathbf{x}_{j-1} - \ldots - (\mathbf{x}_1, \mathbf{y}_j)\mathbf{x}_1$$

and

$$\mathbf{x}_j = \frac{\mathbf{z}_j}{\|\mathbf{z}_j\|} \qquad (\text{for } j = 1, 2, \ldots, n).$$

Results in a set $\{\mathbf{x}_1, \mathbf{x}_2, \ldots, \mathbf{x}_n\}$ of orthonormal vectors.

Example 2.1
Use the Gram–Schmidt process to construct an orthonormal basis for the three-dimensional space $V^{(3)}$, given a basis

$$\{\mathbf{y}_1 = [1\ 0\ 0],\quad \mathbf{y}_2 = [1\ 0\ -1],\quad \mathbf{y}_3 = [2\ 1\ -1]\}.$$

Solution

$$\mathbf{z}_1 = \begin{bmatrix} 1 \\ 0 \\ 0 \end{bmatrix} \quad \text{also} \quad \mathbf{x}_1 = \begin{bmatrix} 1 \\ 0 \\ 0 \end{bmatrix}$$

$$\mathbf{z}_2 = \begin{bmatrix} 1 \\ 0 \\ -1 \end{bmatrix} - 1 \begin{bmatrix} 1 \\ 0 \\ 0 \end{bmatrix} = \begin{bmatrix} 0 \\ 0 \\ -1 \end{bmatrix} \quad \text{also} \quad \mathbf{x}_2 = \begin{bmatrix} 0 \\ 0 \\ -1 \end{bmatrix}$$

$$\mathbf{z}_3 = \begin{bmatrix} 2 \\ 1 \\ -1 \end{bmatrix} - 1 \begin{bmatrix} 0 \\ 0 \\ -1 \end{bmatrix} - 2 \begin{bmatrix} 1 \\ 0 \\ 0 \end{bmatrix} = \begin{bmatrix} 0 \\ 1 \\ 0 \end{bmatrix} \quad \text{also} \quad \mathbf{x}_3 = \begin{bmatrix} 0 \\ 1 \\ 0 \end{bmatrix}.$$

The Gram–Schmidt process is a useful tool in the construction of unitary matrices, as explained below.

Let $A = \{\mathbf{x}_1, \mathbf{x}_2, \ldots, \mathbf{x}_r\}$ be a set of r orthonormal vectors of order $n\ (r < n)$.

Let $B = \{\mathbf{u}_1, \mathbf{u}_2, \ldots, \mathbf{u}_n\}$ be a set of n linearly independent vectors of order n.

Using these two sets of vectors, we wish to construct a unitary matrix U having the (partitioned) form

$$U = [\mathbf{x}_1 \mathbf{x}_2 \ldots \mathbf{x}_r \mathbf{y}_1 \mathbf{y}_2 \ldots \mathbf{y}_{n-r}]. \tag{2.12}$$

The construction proceeds as follows.

Consider the $(n + r)$ vectors

$$\mathbf{x}_1, \mathbf{x}_2, \ldots, \mathbf{x}_r,\quad \mathbf{u}_1, \mathbf{u}_2, \ldots, \mathbf{u}_n.$$

By a simple process (see [B7] p. 56, theorem 2.5) we can reduce this to a set of n linearly independent vectors of the form

$$\mathbf{x}_1, \mathbf{x}_2, \ldots, \mathbf{x}_r, \mathbf{w}_1, \mathbf{w}_2, \ldots, \mathbf{w}_{n-r}$$

where $\mathbf{w}_1, \mathbf{w}_2, \ldots, \mathbf{w}_{n-r}$ are $(n - r)$ vectors from the set B. Now applying the Gram–Schmidt process to the above vectors we finally obtain the set

$$\{\mathbf{x}_1, \mathbf{x}_2, \ldots, \mathbf{x}_r, \ \mathbf{y}_1, \mathbf{y}_2, \ldots, \mathbf{y}_{n-r}\}.$$

This is a set of n orthonormal vectors and so form the columns of the matrix U in (2.12).

It can be shown that every square matrix can be reduced to a triangular form by a similarity transformation. We now consider a particularly interesting example of this.

The method which is discussed below necessitates the calculation of eigenvalues and the corresponding eigenvectors of the matrices involved, nevertheless it is great theoretical importance.

Theorem 2.1

Every matrix $A(n \times n)$ can be reduced by a unitary transformation to a (upper) triangular form matrix.

Note: This result is sometimes called the *Schur triangularisation theorem.*

Proof

Our aim is to show that for every matrix A, there exists a unitary matrix U such that

$$U^*AU = T, \tag{2.13}$$

where T is an upper triangular matrix.

Let λ_1 be an eigenvalue of A and $\mathbf{x}_1$ the associated normalised eigenvector. Choosing a set of n linearly independent vectors which includes $\mathbf{x}_1$, we use the above procedure to construct a unitary matrix Q with $\mathbf{x}_1$ for its first column

$$Q = [\mathbf{x}_1 \mathbf{u}_2 \ldots \mathbf{u}_n] = [\mathbf{x}_1 \;\vdots\; Q_1]$$

where Q_1 is the matrix having columns

$$\mathbf{u}_2, \mathbf{u}_3, \ldots, \mathbf{u}_n.$$

and is such that $Q_1^* \mathbf{x}_1 = \mathbf{0}$.

We now have

$$Q^*AQ = \begin{bmatrix} \mathbf{x}_1^* \\ \hdashline Q_1^* \end{bmatrix} A[\mathbf{x}_1 \;\vdots\; Q_1]$$

$$= \begin{bmatrix} \mathbf{x}_1^* \\ \hdashline Q_1^* \end{bmatrix} [\lambda_1 \mathbf{x}_1 \;\vdots\; AQ_1]$$

$$= \left[\begin{array}{c:c} \lambda_1 \mathbf{x}_1^* \mathbf{x}_1 & \mathbf{x}_1^* AQ_1 \\ \hdashline \lambda_1 Q_1^* \mathbf{x}_1 & Q_1^* AQ_1 \end{array}\right] = \left[\begin{array}{c:c} \lambda_1 & b_1^* \\ \hdashline 0 & A_1 \end{array}\right] \tag{2.14}$$

where $\mathbf{b}_1^* = \mathbf{x}_1^* A Q_1$ and $A_1 = Q_1^* A Q_1$ is a square matrix of order $(n-1)$.

The proof now proceeds by induction. For $n = 2$, the (partitioned) matrix on the right-hand side of (2.14) is in the upper triangular form and Q is a unitary matrix by construction, hence the theorem is true. Assume the theorem is true for a square matrix of order $(n-1)$. Then there exists a unitary matrix W of order $(n-1)$ such that

$$W^* A_1 W = T_1$$

where T_1 is an upper triangular matrix

The matrix

$$V = \left[\begin{array}{c|c} 1 & 0 \\ \hline 0 & W \end{array}\right]$$

is unitary and

$$V^* Q^* A Q V = \left[\begin{array}{c|c} 1 & 0 \\ \hline 0 & W^* \end{array}\right] \left[\begin{array}{c|c} \lambda_1 & \mathbf{b}_1 \\ \hline 0 & A_1 \end{array}\right] \left[\begin{array}{c|c} 1 & 0 \\ \hline 0 & W \end{array}\right]$$

or

$$U^* A U = \left[\begin{array}{c|c} \lambda_1 & \mathbf{b}_1^* W \\ \hline 0 & W^* A_1 W \end{array}\right] = \left[\begin{array}{c|c} \lambda_1 & \mathbf{b}_1^* W \\ \hline 0 & T_1 \end{array}\right]$$

where $U = QV$ is a unitary matrix since both Q and V are unitary, and the matrix on the right-hand side is in the upper triangular form. This proves the theorem.

Example 2.2

Obtain a unitary transformation which transforms the matrix

$$A = \begin{bmatrix} 7 & 4 & -4 \\ 4 & 7 & -4 \\ -1 & -1 & 4 \end{bmatrix}$$

into a triangular form.

Solution

Since the elements of the matrix A are all real, for 'unitary' read 'orthogonal' transformation.

The characteristic equation of A is

$$|\lambda I - A| = (\lambda - 3)^2 \, (\lambda - 12) = 0$$

corresponding to $\lambda = 3$, a normalised eigenvector is

$$\mathbf{x} = \left[\frac{1}{\sqrt{2}}\,0\,\frac{1}{\sqrt{2}}\right]'.$$

Using the Gram–Schmidt process on vectors

$$\begin{bmatrix} \frac{1}{\sqrt{2}} \\ 0 \\ \frac{1}{\sqrt{2}} \end{bmatrix}, \begin{bmatrix} 0 \\ 1 \\ 0 \end{bmatrix} \text{ and } \begin{bmatrix} 0 \\ 0 \\ 1 \end{bmatrix} \text{ (say)}$$

we obtain the orthogronal matrix

$$Q = \begin{bmatrix} \frac{1}{\sqrt{2}} & 0 & -\frac{1}{\sqrt{2}} \\ 0 & 1 & 0 \\ \frac{1}{\sqrt{2}} & 0 & \frac{1}{\sqrt{2}} \end{bmatrix}$$

and hence

$$Q^{-1}AQ = \left[\begin{array}{c:cc} 3 & \frac{3}{\sqrt{2}} & -3 \\ \hdashline 0 & 7 & -\frac{8}{\sqrt{2}} \\ 0 & -\frac{5}{\sqrt{2}} & 8 \end{array}\right]$$

so that

$$A_1 = \begin{bmatrix} 7 & -\frac{8}{\sqrt{2}} \\ -\frac{5}{\sqrt{2}} & 8 \end{bmatrix}.$$

The eigenvalues of A_1 are $\lambda = 3$ and $\lambda = 12$ (the remaining eigenvalues of A).

Corresponding to $\lambda = 3$, the normalised eigenvector is

$$\mathbf{x} = \left[\frac{2}{\sqrt{6}}\ \frac{\sqrt{2}}{\sqrt{6}}\right]'.$$

Using the Gram–Schmidt process on the vectors

$$\begin{bmatrix} \dfrac{2}{\sqrt{6}} \\ \dfrac{\sqrt{2}}{\sqrt{6}} \end{bmatrix} \quad \text{and} \quad \begin{bmatrix} 1 \\ 0 \end{bmatrix} \quad \text{(say)}$$

we obtain the orthogonal matrix

$$W = \begin{bmatrix} \dfrac{2}{\sqrt{6}} & \dfrac{1}{\sqrt{3}} \\ \dfrac{\sqrt{2}}{\sqrt{6}} & -\dfrac{\sqrt{2}}{\sqrt{3}} \end{bmatrix}.$$

It follows that

$$V = \left[\begin{array}{c:c} 1 & 0 \\ \hdashline 0 & W \end{array}\right] = \begin{bmatrix} 1 & 0 & 0 \\ 0 & \dfrac{2}{\sqrt{6}} & \dfrac{1}{\sqrt{3}} \\ 0 & \dfrac{\sqrt{2}}{\sqrt{6}} & -\dfrac{\sqrt{2}}{\sqrt{3}} \end{bmatrix}$$

so that

$$U = QV = \begin{bmatrix} \dfrac{1}{\sqrt{2}} & -\dfrac{1}{\sqrt{6}} & \dfrac{1}{\sqrt{3}} \\ 0 & \dfrac{2}{\sqrt{6}} & \dfrac{1}{\sqrt{3}} \\ \dfrac{1}{\sqrt{2}} & \dfrac{1}{\sqrt{6}} & -\dfrac{1}{\sqrt{3}} \end{bmatrix}$$

This is the orthogonal matrix we have been asked to determine.

Indeed, it is simple to verify that

$$U^*AU = \begin{bmatrix} 3 & 0 & \dfrac{3\sqrt{3}}{\sqrt{2}} \\ 0 & 3 & \dfrac{3}{\sqrt{2}} \\ 0 & 0 & 12 \end{bmatrix} = T.$$

The upper triangular matrix T in (2.13) is sometimes called a *Schur canonical form* of A. A can be written as the product

$$A = UTU^* \tag{2.15}$$

which is known as a *Schur decomposition of A*. Notice that the eigenvalues of A are displayed along the principal diagonal of T.

2.3 HERMITIAN MATRICES

Associated with many practical systems are symmetric matrices. These matrices have various properties which are of great interest in the development of Matrix Theory and are of immense help in solving many practical problems. In the more general case, when it is convenient to consider a matrix with complex elements; or as we say, a complex matrix, it is found that the symmetric complex matrix does *not* possess a number of those properties. On the other hand, the complex matrix A which satifies the relation

$$A^* = A \tag{2.16}$$

is called *Hermitian*. It not only possesses these properties but when the elements of A are all real (2.16) becomes

$$A' = A$$

showing that A is a symmetric matrix.

An example of a Hermitian matrix is

$$A = \begin{bmatrix} 1 & 2-i \\ 2+i & -2 \end{bmatrix}$$

Since the (i, j)th element of a Hermitian matrix satisfies

$$a_{ij} = \bar{a}_{ji},$$

it follows that

$$a_{ii} = \bar{a}_{ii}$$

so that the diagonal elements of such a matrix must all be real.

We next discuss some of the properties of Hermitian matrices.

(1) If A is Hermitian, then

$$(A\mathbf{x}, \mathbf{y}) = (\mathbf{x}, A\mathbf{y}) \tag{2.17}$$

Proof

By (1.25)

$$(A\mathbf{x}, \mathbf{y}) = (A\mathbf{x})^*\mathbf{y} = \mathbf{x}^*A^*\mathbf{y}$$

$$= \mathbf{x}^*(A\mathbf{y}) \qquad \text{by (2.16)}$$

$$= (\mathbf{x}, A\mathbf{y}). \qquad \text{by (1.25)}$$

(2) If A is Hermitian and P is arbitrary, then

$$P^*AP \text{ is Hermitian} \tag{2.18}$$

Proof

$$\begin{aligned}(P^*AP)^* &= P^*A^*P && \text{since } (P^*)^* = P \\ &= P^*AP && \text{by (2.16).}\end{aligned}$$

The most important property follows.

(3) (i) The eigenvalues of a Hermitian matrix are real.
(ii) The eigenvectors corresponding to distinct eigenvalues of a Hermitian matrix are orthogonal.

Proof

(i) Given a Hermitian matrix A, having an eigenvalue λ and a corresponding eigenvector $\mathbf{x}$, then

$$A\mathbf{x} = \lambda\mathbf{x}$$

and

$$(A\mathbf{x}, \mathbf{x}) = (\lambda\mathbf{x}, \mathbf{x}) = \lambda(\mathbf{x}, \mathbf{x}).$$

It follows that

$$\lambda = \frac{(A\mathbf{x}, \mathbf{x})}{(\mathbf{x}, \mathbf{x})}. \tag{2.19}$$

The denominator of (2.19) is real (see (1.27)). The numerator is also real. Indeed, let a bar over an expression denote its complex conjugate (as usual), then

$$\begin{aligned}\overline{(A\mathbf{x}, \mathbf{x})} &= \overline{(A\mathbf{x})^*\mathbf{x}} = \overline{\mathbf{x}^*A^*\mathbf{x}} = \overline{\mathbf{x}^*A\mathbf{x}} \\ &= \overline{(\mathbf{x}, A\mathbf{x})} = (A\mathbf{x}, \mathbf{x}) && \text{by (1.26).}\end{aligned}$$

It follows that λ is the ratio of two real expressions and so is also real.

Note: Given any $\mathbf{x}$, the expression on the right-hand side of (2.19) that is

$$\psi(\mathbf{x}) = \frac{(A\mathbf{x}, \mathbf{x})}{(\mathbf{x}, \mathbf{x})} \tag{2.20}$$

is called the *Rayleigh's quotient.*

(ii) Let $\mathbf{x}$ and $\mathbf{y}$ be the eigenvectors corresponding to distinct eigenvalues λ and μ of A, then

$$A\mathbf{x} = \lambda\mathbf{x} \quad \text{and} \quad A\mathbf{y} = \mu\mathbf{y}.$$

Also

$$(A\mathbf{x}, \mathbf{y}) = \lambda(\mathbf{x}, \mathbf{y})$$

and

$$(\mathbf{x}, A\mathbf{y}) = \mu(\mathbf{x}, \mathbf{y}) \qquad \text{(since } \mu \text{ is real).}$$

Hence

$$\lambda(\mathbf{x}, \mathbf{y}) = \mu(\mathbf{x}, \mathbf{y}) \qquad \text{by (2.17).}$$

But $\lambda \neq \mu$, so that $(\mathbf{x}, \mathbf{y}) = 0$. The result follows.

From this last property, it follows that if the Hermitian matrix A has distinct eigenvalues $\lambda_1, \lambda_2, \ldots, \lambda_n$ the corresponding normalised eigenvectors form an orthonormal set.

Hence the *modal* matrix U, whose columns are the normalised eigenvectors, is *unitary*. It follows that

$$U^{-1}AU = U^*AU = \Lambda$$

where $\Lambda = \operatorname{diag}\{\lambda_1, \lambda_2, \ldots, \lambda_n\}$.

This result assumed that the Hermitian matrix A has distinct eigenvalues. It is a remarkable fact that a Hermitian matrix can always be diagonalised even when the eigenvalues $\lambda_1, \lambda_2, \ldots, \lambda_n$ are not distinct.

Theorem 2.2

If $A(n \times n)$ is a Hermitian matrix, having eigenvalues $\lambda_1, \lambda_2, \ldots, \lambda_n$ (not necessarily distinct) then:

(1) There exists a unitary matrix U such that

$$U^*AU = \Lambda \tag{2.21}$$

where $\Lambda = \operatorname{diag}\{\lambda_1, \lambda_2, \ldots, \lambda_n\}$.

(2) The n eigenvectors of A are linearly independent.

(3) If A has an eigenvalue λ_1 (say) of multiplicity r, there is a set of r orthonormal eigenvectors corresponding to λ_1.

Proof

(1) We proceed as in Theorem 2.1 and obtain (eqn. (2.14))

$$U^*AU = \left[\begin{array}{c:c} \lambda_1 & \mathbf{b}_1^* \\ \hdashline 0 & A_1 \end{array}\right]$$

But by (2.18) U^*AU is Hermitian, so that $\mathbf{b}_1^* = \mathbf{0}$.

Hence the theorem is true for $n = 2$. The proof by induction is now similar to the one used for Theorem 2.1.

(2) Since

$$U^*AU = \Lambda$$

$$AU = U\Lambda \qquad \text{by (2.2)}$$

so that

$$A\mathbf{x}_j = \lambda_j \mathbf{x}_j \qquad (j = 1, 2, \ldots, n)$$

where

$$U = [\mathbf{x}_1 \mathbf{x}_2 \ldots \mathbf{x}_n].$$

So $\mathbf{x}_j (j = 1, 2, \ldots, n)$ are the (normalised) eigenvectors of A and they form a set of orthonormal vectors being the columns of the unitary matrix U. They are therefore linearly independent.

(3) The result follows from (2). The first r elements along the principal diagonal of Λ are all λ_1. The corresponding eigenvectors are the first r columns of U, that is $\mathbf{x}_1, \mathbf{x}_2, \ldots, \mathbf{x}_r$. This is an orthonormal set of vectors.

Notice that there is no loss in generality in assuming that the r-fold eigenvalue of A is λ_1. It could be any of the eigenvalues. We consider a simple illustrative example.

Example 2.3

Find the unitary matrix U such that

$$U^*AU = \Lambda$$

for

$$A = \begin{bmatrix} 1 & 0 & 0 & 0 \\ 0 & 1 & 0 & 0 \\ 0 & 0 & 0 & 1+i \\ 0 & 0 & 1-i & 0 \end{bmatrix}$$

Solution

A is a Hermitian matrix. The characteristic equation of A is

$$C(\lambda) = (\lambda^2 - 2)(\lambda - 1)^2.$$

The eigenvalues of A are $\lambda_1 = \sqrt{2}$, $\lambda_2 = -\sqrt{2}$, $\lambda_3 = 1 = \lambda_4$. Corresponding to $\lambda = \sqrt{2}$, the normalised eigenvector is

$$\mathbf{x} = \left[0 \; 0 \; \frac{1}{\sqrt{2}} \; \frac{1-i}{2}\right]'.$$

This problem is too simple to necessitate the use of the Gram–Schmidt process. By inspection, using (1.28) we obtain an orthonormal set of vectors and construct a unitary matrix U.

$$U = \begin{bmatrix} 0 & 0 & 1 & 0 \\ 0 & 0 & 0 & 1 \\ \frac{1}{\sqrt{2}} & \frac{1+i}{2} & 0 & 0 \\ \frac{1-i}{2} & -\frac{1}{\sqrt{2}} & 0 & 0 \end{bmatrix}$$

It is simple to verify that

$$U^*AU = \text{diag}\{\sqrt{2}, -\sqrt{2}, 1, 1\}.$$

We have seen that the eigenvalues $\{\lambda_i\}$ of a Hermitian matrix of order n are all real. We can therefore order the eigenvalues in a manner so that

$$\lambda_1 \geqslant \lambda_2 \geqslant \ldots \geqslant \lambda_n. \tag{2.22}$$

Further, the normalised eigenvectors $\mathbf{x}_1, \mathbf{x}_1, \ldots, \mathbf{x}_n$ associated with these eigenvalues form an orthonormal set.

It follows by (2.19) that

$$\lambda_1 = (A\mathbf{x}_1, \mathbf{x}_1) \geqslant \lambda_2 = (A\mathbf{x}_2, \mathbf{x}_2) \geqslant \ldots \geqslant \lambda_n = (A\mathbf{x}_n, \mathbf{x}_n) \tag{2.23}$$

This last set of inequalities, suggests that it may be possible to evaluate some eigenvalue by considering

$$(A\mathbf{x}, \mathbf{x})$$

for all $\mathbf{x}$ such that $\|\mathbf{x}\| = 1$.

Since the eigenvectors form an orthonormal basis for a n-dimensional vector space $V^{(n)}$, then for any $\mathbf{x} \in V^{(n)}$

$$\mathbf{x} = \alpha_1 \mathbf{x}_1 + \alpha_2 \mathbf{x}_2 + \ldots + \alpha_n \mathbf{x}_n = \sum_{i=1}^{n} \alpha_i \mathbf{x}_i$$

for some scalars $\alpha_1, \alpha_2, \ldots, \alpha_n$.

Also

$$\begin{aligned}(\mathbf{x}, \mathbf{x}) = 1 &= \left(\sum_{i=1}^{n} \alpha_i \mathbf{x}_i, \sum_{j=1}^{n} \alpha_j \mathbf{x}_j \right) \\ &= \sum_{i,j} \bar{\alpha}_i \alpha_j (\mathbf{x}_i, \mathbf{x}_j) \\ &= \sum_{i,j} \bar{\alpha}_i \alpha_j \delta_{ij} \\ &= \sum_i |\alpha_i|^2.\end{aligned}$$

Since

$$A\mathbf{x} = A\sum \alpha_i \mathbf{x}_i = \sum \alpha_i \lambda_i \mathbf{x}_i,$$

it follows that

$$(A\mathbf{x}, \mathbf{x}) = \left(\sum_i \alpha_i \lambda_i \mathbf{x}_i, \sum_j \alpha_j \mathbf{x}_j\right)$$

$$= \sum \lambda_i |\alpha_i|^2$$

so that by (2.22) and the above

$$(A\mathbf{x}, \mathbf{x}) \leqslant \lambda_1[|\alpha_1|^2 + |\alpha_2|^2 + \ldots + |\alpha_n|^2] = \lambda_1$$

and

$$(A\mathbf{x}, \mathbf{x}) = \lambda_1 \text{ if } \lambda_1 = \lambda_2 \ldots = \lambda_n \text{ or if } \mathbf{x} = \mathbf{x}_1$$

where $\mathbf{x}_1$ is the eigenvector of A corresponding to the largest eigenvalue λ_1 of A.

So it has been shown that

$$(A\mathbf{x}, \mathbf{x}) \leqslant \lambda_1$$

and $(A\mathbf{x}, \mathbf{x}) = \lambda_1$ for some value of $\mathbf{x}$, which is the eigenvector of A corresponding to the eigenvalue λ_1. In other words:

if A is a Hermitian matrix and λ_1 the largest eigenvalue, then

$$\lambda_1 = \max_{\|\mathbf{x}\|=1} (A\mathbf{x}, \mathbf{x}). \tag{2.24}$$

In (2.24) we are confined to $\mathbf{x}$ such that $\|x\| = 1$. To generalise further we now consider any $\mathbf{y} \neq \mathbf{0}$ such that

$$\|\mathbf{y}\| = \alpha > 0 \quad \text{(say)}.$$

Then

$$\frac{(A\mathbf{y}, \mathbf{y})}{(\mathbf{y}, \mathbf{y})} = \frac{1}{\alpha^2}(A\mathbf{y}, \mathbf{y}) \tag{2.25}$$

$$= \left(A\frac{\mathbf{y}}{\alpha}, \frac{\mathbf{y}}{\alpha}\right) = (A\mathbf{x}, \mathbf{x})$$

where

$$\|\mathbf{x}\| = 1.$$

But (2.25) is the Rayleigh's quotient defined in (2.20), which is shown to equal $(A\mathbf{x}, \mathbf{x})$ for some $\mathbf{x}$ such that $\|\mathbf{x}\| = 1$. Conversely, since $(\mathbf{x}, \mathbf{x}) = 1$, then

$$(A\mathbf{x}, \mathbf{x}) = \frac{(A\mathbf{x}, \mathbf{x})}{(\mathbf{x}, \mathbf{x})}.$$

It follows that

$$\lambda_1 = \max_{\mathbf{y}\neq\mathbf{0}} \frac{(A\mathbf{y},\mathbf{y})}{(\mathbf{y},\mathbf{y})} \tag{2.26}$$

where λ_1 is the largest of the eigenvalues of A.

The result of equation (2.26) is sometimes referred to as *Rayleigh's principle.*

Example 2.4

Given

$$A = \begin{bmatrix} 1 & 2 \\ 2 & -2 \end{bmatrix},$$

and vectors $\mathbf{y} \neq \mathbf{0}$, show that

$$\frac{(A\mathbf{y},\mathbf{y})}{(\mathbf{y},\mathbf{y})} \leqslant 2.$$

Solution

We consider various vectors; say

$$\mathbf{y} = \begin{bmatrix} 1 \\ 1 \end{bmatrix}, \begin{bmatrix} 1 \\ -1 \end{bmatrix}, \begin{bmatrix} 2 \\ 3 \end{bmatrix}, \begin{bmatrix} 2 \\ 1 \end{bmatrix}$$

and evaluate $(A\mathbf{y},\mathbf{y})/(\mathbf{y},\mathbf{y})$, to obtain the Rayleigh's quotient

$$\psi\begin{bmatrix} 1 \\ 1 \end{bmatrix} = 1.5, \quad \psi\begin{bmatrix} 1 \\ -1 \end{bmatrix} = -2.5, \quad \psi\begin{bmatrix} 2 \\ 3 \end{bmatrix} = 0.77, \quad \psi\begin{bmatrix} 2 \\ 1 \end{bmatrix} = 2.$$

From these results we conclude that

$$\lambda_1 \leqslant 2.$$

In fact the eigenvalues of A are

$$\lambda_1 = 2 > \lambda_2 = -3.$$

The (non-normalised) eigenvector corresponding to $\lambda_1 = 2$ is $\mathbf{y} = [2\ 1]'$.

We next consider very interesting results known as *Cauchy's inequalities* about the eigenvalues of a principal submatrix of a Hermitian matrix.

Theorem 2.3

Let $A(n \times n)$ be a Hermitian matrix with eigenvalues

$$\lambda_1 \geqslant \lambda_2 \geqslant \ldots \geqslant \lambda_n.$$

Let $B(n-1 \times n-1) = A(\alpha)$, $\alpha = \{r\}$ (see Section 1.3) be a principal submatrix of A with eigenvalues

$$\mu_1 \geqslant \mu_2 \geqslant \ldots \geqslant \mu_{n-1}.$$

Then

$$\lambda_1 \geqslant \mu_1 \geqslant \lambda_2 \geqslant \mu_2 \geqslant \ldots \geqslant \mu_{n-1} \geqslant \lambda_n.$$

This is called the *interlacing property* of eigenvalues.

Proof

By Theorem 2.2, equation (2.21) we can write

$$A = U\Lambda U^*$$

where $U(n \times n)$ is a unitary matrix and

$$\Lambda = \operatorname{diag}\{\lambda_1, \lambda_2, \ldots, \lambda_n\}.$$

Writing (1.13) in the form

$$\frac{\operatorname{adj} g}{C(\lambda)} = g^{-1}$$

and taking

$$g(\lambda) = \lambda I - A,$$

we have

$$\frac{\operatorname{adj}\,[\lambda I - A]}{C(\lambda)} = [\lambda I - A]^{-1} = U^*[\lambda I - \Lambda]^{-1}U \tag{2.27}$$

(remember, $U^{-1} = U^*$) where $C(\lambda) = |\lambda I - A|$.

To consider the form of this matrix we begin by examining the right-hand side of (2.27), for simplicity we take A of order 2×2. Then

$$U^*[\lambda I - \Lambda]^{-1}U = \begin{bmatrix} \bar{u}_{11} & \bar{u}_{21} \\ \bar{u}_{12} & \bar{u}_{22} \end{bmatrix} \begin{bmatrix} \dfrac{1}{\lambda - \lambda_1} & 0 \\ 0 & \dfrac{1}{\lambda - \lambda_2} \end{bmatrix} \begin{bmatrix} u_{11} & u_{12} \\ u_{21} & u_{22} \end{bmatrix}$$

$$= \begin{bmatrix} \dfrac{|u_{11}|^2}{\lambda - \lambda_1} + \dfrac{|u_{21}|^2}{\lambda - \lambda_2} & 0 \\ 0 & \dfrac{|u_{12}|^2}{\lambda - \lambda_1} + \dfrac{|u_{22}|^2}{\lambda - \lambda_2} \end{bmatrix}.$$

This matrix is diagonal, and in the general case when A is of order $(n \times n)$, the rth diagonal element of the matrix on the right-hand side of (2.27) is

$$\sum_{j=1}^{n} \frac{|u_{jr}|^2}{\lambda - \lambda_j}.$$

Now consider the left-hand side of (2.27). The adjoint matrix was defined in Section 1.8. The rth diagonal element of adj $[\lambda I - A]$ is the cofactor of the (r, r)th element in $[\lambda I - A]$.

But $B = A(\alpha)$ is the submatrix obtained by deleting the rth row and the rth column from A.

Hence the (r, r)th element of the left-hand side of (2.27) is

$$\frac{|\lambda I - B|}{C(\lambda)}.$$

It follows that

$$\frac{|\lambda I - B|}{C(\lambda)} = \sum_{j=1}^{n} \frac{|u_{rj}|^2}{\lambda - \lambda_j}. \tag{2.28}$$

We next investigate the behaviour of (2.28) as λ increases from one discontinuity at $\lambda = \lambda_k$ to the neighbouring discontinuity at $\lambda = \lambda_{k-1}$ (say). First assume that $\lambda_{k-1} > \lambda_k$. At $\lambda = \lambda_k + \epsilon$ (where $\epsilon > 0$ is very small) the right-hand side of (2.28) approximately equals

$$\frac{|u_{rk}|^2}{\epsilon}$$

which is large and positive. For $\lambda = \lambda_{k-1} - \epsilon$, the right-hand side of (2.28) approximately equals

$$-\frac{|u_{r,k-1}|^2}{\epsilon}$$

which is large and negative. It follows that (2.28) is monotonically decreasing at points of continuity, and has a zero for $\lambda = \mu_k$ where

$$\lambda_{k-1} \geqslant \mu_k \geqslant \lambda_k.$$

From (2.28) we see that this zero $\lambda = \mu_k$ is an eigenvalue of B. If all the eigenvalues $\lambda_1, \lambda_2, \ldots, \lambda_n$ of A are distinct, then the theorem follows immediately on considering (2.28) for $r = 1, 2, \ldots, n$. On the other hand, if A has multiple eigenvalues, or some $|u_{rj}|^2 = 0$, then it can be shown (for example, by taking appropriate limiting procedures) that some eigenvalues of A and B coincide, but the interlacing property holds.

Example 2.5

Verify Theorem 2.3 for the principal submatrices of

$$A = \begin{bmatrix} 1 & 0 & -4 \\ 0 & 5 & 4 \\ -4 & 4 & 3 \end{bmatrix}.$$

Solution
The characteristic polynomial $C(\lambda)$ of A is

$$C(\lambda) = (\lambda + 3)(\lambda - 3)(\lambda - 9).$$

hence

$$\lambda_1 = 9, \ \lambda_2 = 3, \ \lambda_3 = -3.$$

For $r = 1$, the submatrix is

$$B = \begin{bmatrix} 5 & 4 \\ 4 & 3 \end{bmatrix}.$$

The corresponding eigenvalues are $\mu_1 = 8.12$ and $\mu_2 = 0.12$. In this case

$$\lambda_1 > \mu_1 > \lambda_2 > \mu_2 > \lambda_3.$$

For $r = 2$

$$B = \begin{bmatrix} 1 & -4 \\ -4 & 3 \end{bmatrix}$$

and $\mu_1 = 5.81$ and $\mu_2 = -1.81$.
For $r = 3$

$$B = \begin{bmatrix} 1 & 0 \\ 0 & 5 \end{bmatrix}$$

and $\mu_1 = 5$ and $\mu_2 = 1$.
Again, in the last two cases, the interlacing property holds. ∎

Closely associated with the Hermitian matrix is the *skew-Hermitian matrix A* of order n defined by

$$A^* = -A \tag{2.29}$$

that is,

$$a_{ij} = -\bar{a}_{ji} \quad \text{for all } i, j.$$

Whereas the eigenvalues of a Hermitian matrix are real, the eigenvalues of a skew-Hermitian matrix are pure imaginary. This and other properties of these matrices will be further investigated in Section 2.4 on Normal Matrices and in Problems for Chapter 2.

2.4 NORMAL MATRICES

The *normal matrices* form a family which includes the types already discussed in this chapter.

A matrix $A(n \times n)$ satisfying the condition:

$$AA^* = A^*A \tag{2.30}$$

is called *normal*.

For example a Hermitian matrix satisfies (2.16), so that it also satisfies (2.30). Similarly a unitary matrix satisfies (2.1) and hence also (2.30).

A basic result characterizing normal matrices is as follows:

Theorem 2.4

A matrix $A(n \times n)$ is normal if and only if its n eigenvectors form an orthonormal set.

Proof

First assume that A has a set of n orthonormal eigenvectors which form the columns of a unitary matrix U (the modal matrix). Then

$$U^*AU = \Lambda$$

where Λ is a diagonal matrix.

Hence

$$A = U\Lambda U^*,$$

so that $A^* = U\Lambda^* U$ (note that $U^{-1} = U^*$).

If follows that

$$AA^* = U\Lambda\Lambda^* U \qquad \text{(by 2.1)}$$

and

$$A^*A = U\Lambda^*\Lambda U \qquad \text{(by 2.1)}.$$

Since $\Lambda\Lambda^* = \Lambda^*\Lambda$ (diagonal matrices commute) it follows that

$$AA^* = A^*A$$

so that A is normal.

Next we assume that A is normal. By Theorem 2.1 there exists a unitary matrix U such that

$$U^*AU = T. \tag{2.31}$$

T is normal, indeed

$$T^*T = U^*A^*UU^*AU = U^*A^*AU \qquad \text{(by 2.1)}$$

and

$$TT^* = U^*AUU^*A^*U = U^*AA^*U \qquad \text{(by 2.1)}$$

hence

$$T^*T = TT^* \qquad \text{(by 2.30)}.$$

Writing this equation in terms of the matrix elements, we have

$$\begin{bmatrix} \bar{t}_{11} & 0 & \dots & 0 \\ \bar{t}_{12} & \bar{t}_{22} & & 0 \\ \vdots & \vdots & & \vdots \\ \bar{t}_{1n} & \bar{t}_{2n} & & \bar{t}_{nn} \end{bmatrix} \begin{bmatrix} t_{11} & t_{12} & \dots & t_{1n} \\ 0 & t_{22} & \dots & t_{2n} \\ \vdots & \vdots & & \vdots \\ 0 & 0 & & t_{nn} \end{bmatrix} = \begin{bmatrix} t_{11} & t_{12} & \dots & t_{1n} \\ 0 & t_{22} & & t_{2n} \\ \vdots & \vdots & & \vdots \\ 0 & 0 & & t_{nn} \end{bmatrix} \begin{bmatrix} \bar{t}_{11} & 0 & \dots & 0 \\ \bar{t}_{12} & \bar{t}_{22} & & 0 \\ \vdots & \vdots & & \vdots \\ \bar{t}_{1n} & \bar{t}_{2n} & & \bar{t}_{nn} \end{bmatrix}$$

so that $t_{rj} = 0$ for $r > j$.

On equating the diagonal elements on both sides, we obtain

$$\bar{t}_{1j}t_{1j} + \bar{t}_{2j}t_{2j} + \dots + \bar{t}_{jj}t_{jj} = t_{jj}\bar{t}_{jj} + t_{j,j+1}\bar{t}_{j,j+1} + \dots + t_{jn}\bar{t}_{jn}$$

or

$$\bar{t}_{jj}t_{jj} + \sum_{r=1}^{j-1} |t_{rj}|^2 = \bar{t}_{jj}t_{jj} + \sum_{r=j+1}^{n} \left| t_{j,r} \right|^2 \quad (j = 1, 2, \dots, n).$$

This shows that

$$t_{rj} = 0 \text{ for } r < j.$$

So all the off-diagonal elements of T are zero, hence T is a diagonal matrix and must be the eigenmatrix Λ of A.

It follows that

$$U^*AU = \Lambda$$

and the unitary matrix U is the modal matrix of A.

The result follows.

The above theorem is equivalent to parts (1) and (2) of Theorem 2.2. This is not surprising since a Hermitian matrix is a normal matrix. Also, it should not be surprising that part (3) of Theorem 2.2 is valid for normal matrices in general. The reader is invited to prove this fact for himself.

If we are given a normal matrix A we can characterise it as unitary, Hermitian, or skew-Hermitian by definitions (2.1), (2.16) or (2.29) respectively. On the other hand, it can be classified in terms of its eigenvalues as follows:

(1) A is unitary iff its eigenvalues have modulus 1.
(2) A is Hermitian iff its eigenvalues are real.
(3) A is skew-Hermitian iff its eigenvalues are pure imaginary.

The proof again follows immediately from Theorem 2.4. It was shown (see eqn. 2.31) that if A is normal then there exists a unitary matrix U such that

$$U^*AU = \Lambda.$$

The results now follow, since A and Λ have the same eigenvalues.

An important result concerning the eigenvalues of a matrix is contained in the following theorem, known as *Schur's inequality*.

Theorem 2.5
Given $A(n \times n) = [a_{ij}]$, having eigenvalues $\lambda_1, \lambda_2, \ldots, \lambda_n$, then

$$\sum_{r=1}^{n} |\lambda_r|^2 = \sum_{i,j} |a_{ij}|^2$$

if A is normal, and

$$\sum_{r=1}^{n} |\lambda_r|^2 < \sum_{i,j} |a_{ij}|^2$$

otherwise.

Proof
By Theorem 2.1, there exists a unitary matrix U such that

$$U^*AU = T$$

where T is an upper triangular matrix (or a diagonal matrix if A is normal). It follows that

$$TT^* = U^*AA^*U$$

and by (1.39)

$$\text{tr}\,(TT^*) = \text{tr}\,(AA^*)$$

or, by (1.38)

$$\sum_{i,j} |t_{ij}|^2 = \sum_{i,j} |a_{ij}|^2.$$

But

$$\sum_{i,j} |t_{ij}|^2 = \sum_{i} |t_{ii}|^2 + \sum_{i<j} |t_{ij}|^2$$

since $t_{ij} = 0$ for $i > j$ for every A. (Also $t_{ij} = 0$ for $i < j$ if A is normal.) Hence

$$\sum_{i} |\lambda_i|^2 + \sum_{i<j} |t_{ij}|^2 = \sum_{i,j} |a_{ij}|^2,$$

so that

$$\sum_{i} |\lambda_i|^2 \leqslant \sum_{i,j} |a_{ij}|^2 \tag{2.32}$$

the equality being valid if A is normal.

Example 2.6
Verify Schur's inequality for the matrices considered in Examples 2.2 and 2.3.

Solution
For the matrix in Example 2.2.

$$\sum |a_{ij}|^2 = 180.$$

$$\lambda_1 = \lambda_2 = 3, \ \lambda_3 = 12,$$

hence

$$\sum |\lambda_i|^2 = 162.$$

In this case

$$\sum |\lambda_i|^2 < \sum |a_{ij}|^2.$$

For the matrix in Example 2.3

$$\sum |a_{ij}|^2 = 6$$

$$\lambda_1 = \sqrt{2}, \quad \lambda_2 = -\sqrt{2}, \quad \lambda_3 = \lambda_4 = 1,$$

hence

$$\sum |\lambda_i|^2 = 6.$$

In this case

$$\sum |\lambda_i|^2 = \sum |a_{ij}|^2$$

as predicted, since the matrix is normal.

PROBLEMS

2.1 The first column of a matrix A of order (3×3) is

$$u_1' = \left(\frac{1}{2}, \frac{1}{2}, \frac{\sqrt{2}}{2}\right).$$

Find the other two columns so that A is orthogonal.

2.2 (1) Given

$$U = \begin{bmatrix} a_1 & a_2 \\ a_3 & a_4 \end{bmatrix}$$

find the conditions for U to be a unitary matrix.

(2) Find an unitary matrix whose first column is

$$\left(\frac{1}{2}, \frac{1}{\sqrt{2}}i, \frac{1}{2}\right)'.$$

2.3 Use the Gram–Schmidt process to construct an orthonormal set of vectors $\{\mathbf{x}_1, \mathbf{x}_2, \mathbf{x}_3\}$ given the set $\{\mathbf{y}_1, \mathbf{y}_2, \mathbf{y}_3\}$, where

$$\mathbf{y}_1 = [1, 1, 1]', \ \mathbf{y}_2 = [1, 2, 3]' \text{ and } \mathbf{y}_3 = [0, 1, -1]'.$$

2.4 Show that if P and Q are unitary matrices then so is PQ.

2.5 Find the unitary matrix U such that

$$U^*AU = T$$

where T is an upper triangular matrix, and

$$A = \begin{bmatrix} -1 & \frac{10}{\sqrt{12}} & \frac{1}{\sqrt{12}} \\ 0 & 2 & -\frac{3}{\sqrt{18}} \\ 0 & -\frac{6}{\sqrt{18}} & 2 \end{bmatrix}.$$

2.6 Find the unitary matrix U such that

$$U^*AU$$

is diagonal, when

$$(1) \quad A = \begin{bmatrix} 2 & 2 & 0 \\ 2 & 2 & 0 \\ 0 & 0 & 1 \end{bmatrix} \quad \text{and} \quad (2) \quad A = \begin{bmatrix} 5 & 4 & -4 \\ 4 & 5 & 4 \\ -4 & 4 & 5 \end{bmatrix}.$$

2.7 $A(n \times n)$ is a Hermitian matrix with eigenvalues

$$\lambda_1 \geqslant \lambda_2 \geqslant \ldots \geqslant \lambda_n$$

and associated orthonormal eigenvectors

$$\mathbf{x}_1, \mathbf{x}_2, \ldots, \mathbf{x}_n.$$

Show that

$$\lambda_n \leqslant \psi(\mathbf{x}) \leqslant \lambda_1$$

where $\psi(\mathbf{x})$ is the Rayleigh quotient associated with A and $\mathbf{x} \in V^n$.

2.8 (1) λ_0 and $\mathbf{x}_0$ is an eigenvalue and the corresponding eigenvector of a Hermitian matrix A. Let

$$\mathbf{y} = \mathbf{x}_0 + \epsilon \mathbf{x}$$

be a vector which can be thought of as an approximation to $\mathbf{x}_0$.

If $\psi(\mathbf{y})$ is the Rayleigh quotient associated with A, show that

$$\psi(\mathbf{y}) = \lambda_0 + |\epsilon|^2 [\psi(\mathbf{x}) - \lambda_2] \frac{(\mathbf{x}, \mathbf{x})}{(\mathbf{y}, \mathbf{y})}.$$

(2) Given

$$A = \begin{bmatrix} 4 & 3i \\ -3i & 2 \end{bmatrix}$$

and $\mathbf{y}' = [1 \; -i]$, an approximation to an eigenvector $\mathbf{x}_0$ of A, use the above formula to find an approximation to the corresponding eigenvalue λ_0. Comment on the result.

2.9 $\lambda_1, \lambda_2, \ldots, \lambda_n$ are the eigenvalues of a matrix $A = [a_{ij}]$ of order $(n \times n)$. Let

$$\mu = \max_{i,j} |a_{ij}|.$$

Show that

(1) $|\det A| \leqslant n^{n/2} \mu^n$

(2) $|\lambda_r| \leqslant n\mu \; (r = 1, 2, \ldots, n)$.

Note Use the arithmetic–geometric mean inequality

$$\left(\prod_{i=1}^{n} a_i \right)^{1/n} \leqslant \frac{1}{n} \sum_{i=1}^{n} a_i.$$

2.10 Show that any complex matrix C can be written as

$$C = H + S$$

where H is Hermitian and S is skew-Hermitian.

2.11 Given Hermitian matrices A and B, show that the product AB is Hermitian if

$$AB = BA.$$

2.12 Verify Cauchy's inequalities (interlacing property of eigenvalues) for each of the principal submatrices B_i of A, when

$$(1) \quad A = \begin{bmatrix} 0 & -i & 0 \\ i & 0 & 0 \\ 0 & 0 & 3 \end{bmatrix} \quad \text{and} \quad (2) \quad A = \begin{bmatrix} 5 & 2 & -2 \\ 2 & 5 & -2 \\ -2 & -2 & 5 \end{bmatrix}.$$

2.13 Show that

(1) if A is of order $(n \times n)$ then

$$AA^* \text{ and } A^*A$$

are Hermitian.

(2) if $\mathbf{x}$ and $\mathbf{y}$ are vectors or order n, then

$$\mathbf{y}\mathbf{x}^* + \mathbf{y}\mathbf{x}^*$$

is Hermitian.

3

Positive Definite Matrices

3.1 INTRODUCTION

There are various reasons, explained below, for including a brief survey of the theory of positive definite matrices in this chapter.

Some of the techniques developed for the study of positive definite matrices are equally useful for consideration of nonnegative matrices. Also various properties of positive definite matrices are used to prove other properties of nonnegative matrices. The purpose of studying nonnegative matrices is due to their importance in stochastic theory and other fields. But equally important are positive definite matrices to the theory of conic sections, optimisation, dynamics of physical systems, least-squares problems and other fields. By discussing both positive definite and nonnegative matrices, it is hoped to satisfy many interests, some of them closely related.

As pointed out, there are many common properties binding these two classes of matrices, and of course there are numerous matrices belonging to both classes, the best known being the unit matrix.

Finally, it is stressed that in this chapter only a very limited number of useful techniques are covered. For example, in the discussion of quadratic forms neither Lagrange's nor Kronecker's reductions are mentioned. These techniques are well covered in other literature (see [B10], [B13], for example).

3.2 QUADRATIC FORMS

We consider an expression of the type

$$q(\mathbf{x}) = \sum_{i,j=1}^{n} a_{ij} x_i x_j \tag{3.1}$$

where the coefficients a_{ij} can be real or complex scalars and $\mathbf{x} = [x_1, x_2, \ldots, x_n]'$. For simplicity, writing (3.1) out in full for $n = 2$, we obtain

$$\begin{aligned} q(\mathbf{x}) &= a_{11}x_1^2 + a_{12}x_1x_2 + a_{21}x_2x_1 + a_{22}x_2^2 \\ &= a_{11}x_1^2 + (a_{12} + a_{21})x_1x_2 + a_{22}x_2^2. \end{aligned}$$

We note that $q(\mathbf{x})$ is a homogeneous polynomial in which the coefficient of the product term x_1x_2 is

$$a_{12} + a_{21}.$$

In the general case the coefficient of x_ix_j $(i \neq j)$ is

$$a_{ij} + a_{ji}.$$

We can write (3.1) in matrix form by introducing the matrix to coefficients

$$A = [a_{ij}]. \tag{3.2}$$

Then

$$q(\mathbf{x}) = (\mathbf{x}, A\mathbf{x}). \tag{3.3}$$

To verify (3.3) for $n = 2$, we write $q(\mathbf{x})$ in the form

$$\begin{aligned} & x_1(a_{11}x_1 + a_{12}x_2) + x_2(a_{21}x_1 + a_{22}x_2) \\ &= [x_1 x_2] \begin{bmatrix} a_{11} & a_{12} \\ a_{21} & a_{22} \end{bmatrix} \begin{bmatrix} x_1 \\ x_2 \end{bmatrix} \\ &= (\mathbf{x}, A\mathbf{x}) \qquad \text{(by (1.25) and (2.5)).} \end{aligned}$$

As far as the definition (3.1) is concerned the coefficients a_{ij} are not interrelated. On the other hand it has been pointed out that the coefficient a_{ij} has no independent role in the formation of $q(\mathbf{x})$. It is the sum $(a_{ij} + a_{ji})$, the coefficient of x_ix_j which is important. This simple observation has important repercussions when considering the form of the matrix of coefficients A. It means that we can *choose* A to be a Hermitian matrix (or symmetric, if the coefficients are real) without loss of generality. This will enable us to use various powerful properties of Hermitian matrices for investigating quadratic forms.

For example, we write the matrix of coefficients of

$$q(\mathbf{x}) = x_1^2 - 2x_2^2 + 3x_3^2 - 5x_1x_2 + 4x_1x_3$$

as

$$A = \begin{bmatrix} 1 & -5/2 & 2 \\ -5/2 & -2 & 0 \\ 2 & 0 & 3 \end{bmatrix}$$

although the same polynomial is obtained if we choose in (3.3)

$$A = \begin{bmatrix} 1 & -2 & 4 \\ -3 & -2 & 0 \\ 0 & 0 & 3 \end{bmatrix},$$

which is not symmetric.

It will be remembered from the discussion of the properties of Hermitian matrices (see proof of property (1) on p. 77) that

$$(A\mathbf{x}, \mathbf{x})$$

is a real expression, when A is Hermitian.

Definition

If in (3.3), A is Hermitian and the components x_i in $\mathbf{x} = [x_1 x_2, \ldots, x_n]'$ are complex variables, then $q(\mathbf{x})$ is called a *Hermitian form*. If A is real and symmetric, then $q(\mathbf{x})$ is called a *real Hermitian form*. If A is real and symmetric and the variable $x_1, x_2, \ldots, x_n$ are real, then $q(\mathbf{x})$ is called a *quadratic form*.

Most of our investigations will concern Hermitian forms, of course the results apply also to the special case of quadratic forms.

3.3 REDUCTIONS TO A CANONICAL FORM

We can express the Hermitian form in terms of variables $y_1, y_2, \ldots, y_n$, the components of a vector $\mathbf{y}$, by applying a linear transformation

$$\mathbf{x} = P\mathbf{y} \quad \text{(say)} \tag{3.4}$$

where P is non-singular.

Then

$$\begin{aligned}(\mathbf{x}, A\mathbf{x}) &= (P\mathbf{y}, AP\mathbf{x}) \\ &= (P^*AP\mathbf{y}, \mathbf{y}) \qquad \text{by (2.5).}\end{aligned}$$

(Also remember that $A = A^*$.)

$$= (\Lambda\mathbf{y}, \mathbf{y}). \tag{3.5}$$

If we choose P to be the modal matrix (choosing normalised eigenvectors) of A then Λ is the eigenvalue matrix of A. In Theorem 2.2, it was proved that P is a unitary matrix.

It should be clear that by applying the linear transformation (3.4), we are choosing a new basis with respect to which the matrix of coefficients (3.2) takes on a particular form. It is in our interest to choose this form as simple as possible, diagonal for example, in which case we say that $(A\mathbf{x}, \mathbf{x})$ is in a *diagonal form.*

Example 3.1

Determine an appropriate transformation

$$\mathbf{x} = U\mathbf{y}$$

so that

$$q(\mathbf{x}) = x_1^2 - 2x_2^2 + x_3^2 - 4x_1x_2$$

is transformed into a diagonal form.

Solution

The (symmetrical) matrix of coefficients is

$$A = \begin{bmatrix} 1 & -2 & 0 \\ -2 & -2 & 0 \\ 0 & 0 & 1 \end{bmatrix}$$

$|\lambda I - A| = (\lambda - 1)(\lambda - 2)(\lambda + 3).$

For

$$\lambda_1 = 1, \quad \mathbf{x}_1 = [0 \;\; 0 \;\; 1]\,.$$

$$\lambda_2 = 2, \quad \mathbf{x}_2 = \begin{bmatrix} \frac{2}{\sqrt{5}} & \frac{-1}{\sqrt{5}} & 0 \end{bmatrix}'$$

$$\lambda_3 = -3, \; \mathbf{x}_3 = \begin{bmatrix} \frac{1}{\sqrt{5}} & \frac{2}{\sqrt{5}} & 0 \end{bmatrix}'.$$

Hence

$$U = \begin{bmatrix} 0 & \frac{2}{\sqrt{5}} & \frac{1}{\sqrt{5}} \\ 0 & \frac{-1}{\sqrt{5}} & \frac{2}{\sqrt{5}} \\ 1 & 0 & 0 \end{bmatrix}.$$

It is easy to verify that

$$U^*AU = \begin{bmatrix} 1 & 0 & 0 \\ 0 & 2 & 0 \\ 0 & 0 & -3 \end{bmatrix} = \Lambda$$

Hence the transformed quadratic form is

$$(\Lambda\mathbf{y}, \mathbf{y}) = y_1^2 + 2y_2^2 - 3y_3^2.$$

In Section 1.10 we have shown that for a matrix A

$$\|A\| \geqslant |\lambda|$$

where λ is any eigenvalue of A.

Example 3.2
Show that the Euclidean norm for a matrix A is

$$\|A\| = \lambda_{\max}(A^*A)$$

where $\lambda_{\max}(A^*A)$ indicates the maximum eigenvalue of the matrix A^*A.

Solution
It has been shown above that if Q is a Hermitian matrix, then there exists a unitary matrix U such that

$$(\mathbf{x}, Q\mathbf{x}) = (\mathbf{y}, \Lambda\mathbf{y}) = \lambda_1 |y_1|^2 + \ldots + \lambda_n |y_n|^2$$

where $\mathbf{x} = U\mathbf{y}$, and $\lambda_1, \lambda_2, \ldots, \lambda_n$ are the eigenvalues of Q.
Hence

$$\lambda_{\min}(|y_1|^2 + \ldots + |y_n|^2) \leqslant (\mathbf{x}, Q\mathbf{x}) \leqslant \lambda_{\max}(|y_1|^2 + \ldots + |y_n|^2)$$

where $\lambda_{\max} = \max\{\lambda_1, \ldots, \lambda_n\}$ and $\lambda_{\min} = \min\{\lambda_1, \ldots, \lambda_n\}$ or

$$\lambda_{\min}(\mathbf{y}, \mathbf{y}) \leqslant (\mathbf{x}, Q\mathbf{x}) \leqslant \lambda_{\max}(\mathbf{y}, \mathbf{y}).$$

But

$$(\mathbf{y}, \mathbf{y}) = (U\mathbf{x}, U\mathbf{x}) = (\mathbf{x}, \mathbf{x})$$

(since U is unitary), hence

$$\lambda_{\min}(\mathbf{x}, \mathbf{x}) \leqslant (\mathbf{x}, Q\mathbf{x}) \leqslant \lambda_{\max}(\mathbf{x}, \mathbf{x}).$$

Also

$$\|A\mathbf{x}\|^2 = (A\mathbf{x}, A\mathbf{x}) = (\mathbf{x}, Q\mathbf{x})$$

where $Q = A^*A$ is a Hermitian matrix.
So that

$$\|A\mathbf{x}\|^2 \leqslant \lambda_{\max}\|\mathbf{x}\|^2,$$

or

$$\frac{\|A\mathbf{x}\|^2}{\|\mathbf{x}\|^2} \leqslant \lambda_{\max}.$$

By (1.33) it follows that

$$\|A\| = \lambda_{\max}(A^*A).$$

3.4 A GEOMETRICAL APPLICATION

A well known application of diagonalisation of a quadratic form is to the classification of conic sections. The problem presents itself in the form of a quadratic equation

$$ax^2 + 2bxy + cy^2 = d. \tag{3.6}$$

The object is to determine the nature of the graph defined by (3.6).

As discussed in the previous section, the matrix of coefficients for the expression on the left-hand side of (3.6) is

$$A = \begin{bmatrix} a & b \\ b & c \end{bmatrix}$$

This can be transformed by

$$\mathbf{x} = U\mathbf{y}$$

into a diagonal form, so that

$$(\Lambda \mathbf{y}, \mathbf{y}) = d$$

where

$$\Lambda = U^* A U = \begin{bmatrix} \lambda_1 & 0 \\ 0 & \lambda_2 \end{bmatrix}.$$

(3.6) now becomes

$$\lambda_1 y_1^2 + \lambda_2 y_2^2 = d \tag{3.7}$$

or

$$\frac{y_1^2}{d/\lambda_1} + \frac{y_2^2}{d/\lambda_2} = 1.$$

Hence

(1) If $d/\lambda_1 > 0$ and $d/\lambda_2 > 0$, (3.6) represents an *ellipse.*
(2) If (1) is true and $\lambda_1 = \lambda_2$, (3.6) represents a *circle*.
(3) If d/λ_1 and d/λ_2 are of opposite sign, (3.6) represents a *hyperbola.*
(4) If one of the eigenvalues is zero, (3.7) shows that the expression represents two parallel lines.

Other special cases are left to the reader to sort out.

3.5 POSITIVE DEFINITE MATRICES

Definition

$q(\mathbf{x}) = (\mathbf{x}, A\mathbf{x})$ is *positive definite*, if and only if $(\mathbf{x}, A\mathbf{x}) > 0$ for all (non-zero) real vectors $\mathbf{x}$. If $(\mathbf{x}, A\mathbf{x}) \geqslant 0$ for all real $\mathbf{x} \neq \mathbf{0}$, $q(\mathbf{x})$ is *positive semidefinite.* If $(\mathbf{x}, A\mathbf{x}) < 0$ for all real $\mathbf{x} \neq \mathbf{0}$, $q(\mathbf{x})$ is *negative definite*.

We shall discuss some of the properties of *positive definite* quadratic forms. Properties of other definite and semidefinite forms in general follow immediately.

Definition
We call a (Hermitian) matrix A *positive definite* if $(\mathbf{x}, A\mathbf{x})$ is positive definite. We denote the set of all positive definite matrices of order $(n \times n)$ by Σ_n.

An example of a positive definite quadratic form is

$$q(\mathbf{x}) = x_1^2 + 2x_2^2 + x_3^2 - 2x_1x_2.$$

On the other hand

$$q(\mathbf{x}) = x_1^2 + x_2^2 + x_3^2 - 2x_1x_2 = (x_1 - x_2)^2 + x_3^2$$

is positive semidefinite, since the non-zero vector

$$\mathbf{x} = [\alpha, \alpha, 0]$$

makes $q(\mathbf{x}) = 0$.

It follows that the matrix

$$\begin{bmatrix} 1 & -1 & 0 \\ -1 & 2 & 0 \\ 0 & 0 & 1 \end{bmatrix}$$

is positive definite, but

$$\begin{bmatrix} 1 & -1 & 0 \\ -1 & 1 & 0 \\ 0 & 0 & 1 \end{bmatrix}$$

is not.

Theorem 3.1
If $A \in \Sigma_n$ and $P(n \times n)$ is non-singular, then $P^*AP \in \Sigma_n$.

Proof
Let

$$\mathbf{x} \neq \mathbf{0},$$

then

$$P\mathbf{x} = \mathbf{y} \neq \mathbf{0}.$$

Since $A \in \Sigma_n$

$$(\mathbf{y}, A\mathbf{y}) > 0.$$

But $(\mathbf{y}, A\mathbf{y}) = (P\mathbf{x}, AP\mathbf{x}) = (\mathbf{x}, P^*AP\mathbf{x})$ (by 2.5).

Hence $P^*AP \in \Sigma_n$.

Theorem 3.2
If $A(n \times n)$ is Hermitian, then $A \in \Sigma_n$ if and only if all the eigenvalues $\lambda_1, \lambda_2, \ldots, \lambda_n$ of A are positive.

Proof
Since A is Hermitian, there exists a unitary matrix U (see Theorem 2.2), such that

$$(\mathbf{x}, A\mathbf{x}) = (\mathbf{y}, \Lambda\mathbf{y}) = \lambda_1 \bar{y}_1 y_1 + \lambda_2 \bar{y}_2 y_2 + \ldots + \lambda_n \bar{y}_n y_n \tag{3.8}$$

where

$$\Lambda = \text{diag}\{\lambda_1, \lambda_2, \ldots, \lambda_n\} = U^* A U.$$

and

$$\mathbf{y} = [y_1 y_2, \ldots, y_n]'.$$

If all $\lambda_i > 0$, then the right-hand side of (3.8) is positive definite.

By Theorem 3.1, it follows that $A \in \Sigma_n$.

If $A \in \Sigma_n$, there corresponds to the (real) eigenvalue λ_i of A the normalised eigenvector $\mathbf{x}_i$ ($i = 1, 2, \ldots, n$).

But

$$(\mathbf{x}, A\mathbf{x}) > 0,$$

in particular

$$(\mathbf{x}_i, A\mathbf{x}_i) > 0$$

so that

$$(\mathbf{x}_i, \lambda_i \mathbf{x}_i) = \lambda_i(\mathbf{x}_i, \mathbf{x}_i) = \lambda_i > 0.$$

Corollary
If $A \in \Sigma_n$, then $|A| > 0$.

Proof
By (1.12), $|A| = \lambda_1, \lambda_2, \ldots, \lambda_n$. The result follows. ■

We next make use of the notation introduced in Section 1.3, for example,

$$A(\alpha)$$

is the square submatrix obtained by deleting the rows and columns of A *not* indicated by the index set α.

Theorem 3.3
If $A \in \Sigma_n$, then $A(\alpha)$ is positive definite for each $\alpha \subseteq N$.

Proof
Let

$$\mathbf{x} = [x_1 x_2, \ldots, x_n]' \neq \mathbf{0},$$

where $x_j = 0$ if $j \notin \alpha$ and x_j is arbitrary if $j \in \alpha$.

For example if $\alpha = \{2, 3\}$, then for $n = 4$,

$$\mathbf{x} = [0 \;\; x_2 \;\; x_3 \;\; 0]'$$

where x_2 and x_3 are arbitrary (except that they cannot *both* be zero).

Let $\mathbf{x}(\alpha)$ be the vector obtained from $\mathbf{x}$ by eliminating the (zero) elements $x_j (j \notin \alpha)$.

In the above example, for $n = 4$ and $\alpha = \{2, 3\}$,

$$\mathbf{x}(\alpha) = [x_2 x_3]'.$$

We can now write (for this choice of $\mathbf{x}$)

$$\mathbf{x}'(\alpha)\, A(\alpha)\, \mathbf{x}(\alpha) = \mathbf{x}'A\mathbf{x} = (\mathbf{x}, A\mathbf{x}) > 0.$$

But $\mathbf{x}(\alpha)$ is arbitary, hence $A(\alpha)$ is positive definite.

Corollary

If $A \in \Sigma_n$, the diagonal elements of A are positive.

Proof

Choose $\alpha = \{i\}$, then $A(\alpha) = a_{ii}$. The result follows on taking successively $i = 1, 2, \ldots, n$. ■

We next consider a theorem which is a fundamental result characterising positive definite matrices.

Theorem 3.4

A necessary and sufficient condition for $A(n \times n) \in \Sigma_n$ where $A = [a_{ij}]$, is that the principal minors $|A(\alpha)|$ for each $\alpha \subseteq N$ are all positive, that is;

$$a_{11} > 0, \quad \begin{vmatrix} a_{11} & a_{12} \\ a_{21} & a_{22} \end{vmatrix} > 0, \ldots, \quad \begin{vmatrix} a_{11} & \ldots & a_{1n} \\ \vdots & & \vdots \\ a_{n1} & \ldots & a_{nn} \end{vmatrix} > 0. \tag{3.9}$$

Proof

First assume that $A \in \Sigma_n$. Then by Theorem 3.3, and the fact that $A(\alpha)$ is Hermitian, $A(\alpha) \in \Sigma_i$ for $i = 1, 2, \ldots, n$ and corresponding $\alpha \subseteq N$. Conditions (3.9) now follow by the corollary to Theorem 3.1.

Nex, assume that (3.9) holds.

We make a slight simplification of notation, and write

$$|A_k|$$

as the leading principal minor of order k.

In other words

$$A_k = A(\alpha) = \begin{bmatrix} a_{11} & \ldots & a_{1k} \\ \vdots & & \vdots \\ a_{k1} & \ldots & a_{kk} \end{bmatrix} \qquad \text{for } \alpha = \{1, 2, \ldots, k\}.$$

The proof is now by induction. For $n = 1$, $|A_1| = a_{11} > 0$, hence $A_1 \in \Sigma_1$. Assume that for $n = k$, $A_k \in \Sigma_k$. Let $\lambda_1 \geqslant \lambda_2 \geqslant \ldots \geqslant \lambda_{k+1}$ be the eigenvalues of A_{k+1} and $\mu_1 \geqslant \mu_2 \ldots \geqslant \mu_k$ be the eigenvalues of A_k. By Cauchy's inequality (Theorem 2.3)

$$\lambda_1 \geqslant \mu_1 \geqslant \lambda_2 \geqslant \mu_2 \geqslant \ldots \geqslant \lambda_k \geqslant \mu_k \geqslant \lambda_{k+1}.$$

Since $A_k \in \Sigma_k$ the eigenvalues $\mu_1, \mu_2, \ldots, \mu_k$ are positive (Theorem 3.2), it follows that $\lambda_1, \lambda_2, \ldots, \lambda_k$ are positive. But by (1.12), the product of eigenvalues, that is

$$\lambda_1 \lambda_2 \ldots \lambda_k \lambda_{k+1} = |A_{k+1}| > 0.$$

So that $\lambda_{k+1} > 0$, hence

$$(\mathbf{x}, A_{k+1}\mathbf{x}) = (\mathbf{y}, \Lambda_{k+1}\mathbf{y}) > 0$$

where

$$\Lambda_{k+1} = \text{diag}\,\{\lambda_1, \lambda_2, \ldots, \lambda_{k+1}\}.$$

or

$$A_{k+1} \in \sum\nolimits_{k+1}.$$

We have proved by induction, that

$$A = A_n \in \sum\nolimits_n. \quad \blacksquare$$

A useful concept associated with a positive definite matrix A is the *square root* of A. We begin by considering a diagonal matrix.

Definition

Given a matrix

$$\Lambda = \text{diag}\,\{\lambda_1, \ldots, \lambda_n\}$$

where

$$\lambda_i > 0 \qquad (i = 1, 2, \ldots, n),$$

then

$$\Lambda^{1/2} = \text{diag}\,\{\lambda_1^{1/2} \ldots, \lambda_n^{1/2}\}.$$

To ensure that $\Lambda^{1/2}$ is uniquely defined, we take the positive roots of the eigenvalues λ_i.

The definition is easily generalised to any positive definite matrix A.

Since (Theorem 2.2)

$$A = U\Lambda U^*$$

we define

$$A^{1/2} = U\Lambda^{1/2}U^* \tag{3.10}$$

Notice that (3.10) implies that

$$A^{1/2}A^{1/2} = U\Lambda^{1/2}U^*U\Lambda^{1/2}U^* = U\Lambda U^* = A.$$

Similarly we define the unique matrix $A^{-(1/2)}$.

$$A^{-(1/2)} = U\Lambda^{-(1/2)}U^* \tag{3.11}$$

where

$$\Lambda^{-(1/2)} = \text{diag}\left\{\lambda_1^{-(1/2)}, \ldots, \lambda_n^{-(1/2)}\right\}.$$

Example 3.3
Find $A^{1/2}$ and $A^{-(1/2)}$, given

$$A = \begin{bmatrix} 8 & -2i \\ 2i & 5 \end{bmatrix}.$$

Solution

$$C(\lambda) = |\lambda I - A| = (\lambda - 4)(\lambda - 9)$$

For

$$\lambda_1 = 4, \quad \mathbf{x}_1 = [1 + i \quad 2 - 2i]'$$

for

$$\lambda_2 = 9, \quad \mathbf{x}_2 = [2 - 2i \quad 1 + i]'.$$

Hence

$$U = \frac{1}{\sqrt{10}}\begin{bmatrix} 1+i & 2-2i \\ 2-2i & 1+i \end{bmatrix}.$$

$$U^*AU = \text{diag}\{4, 9\} = \Lambda,$$

so that

$$\Lambda^{1/2} = \text{diag}\{2, 3\} \quad \text{and} \quad \Lambda^{-(1/2)} = \text{diag}\left\{\frac{1}{2}, \frac{1}{3}\right\}.$$

It follows that

$$A^{1/2} = U\Lambda^{1/2}U^* = \frac{1}{5}\begin{bmatrix} 14 & -2i \\ 2i & 11 \end{bmatrix}$$

and

$$A^{-(1/2)} = U\Lambda^{-(1/2)}U^* = \frac{1}{30}\begin{bmatrix} 11 & 2i \\ -2i & 14 \end{bmatrix}.$$

By simple matrix multiplication we can check (for example) that

$$A^{1/2}A^{-(1/2)} = I$$

and

$$A^{-(1/2)}A^{-(1/2)} = A^{-1}$$

The above definition can be further generalised as follows:

Definition
If $A \in \Sigma_n$, then

$$A^{1/r}$$

is the matrix B such that

$$B^r = A.$$

As above we define

$$A^{1/r} = U\Lambda^{1/r}U^* \tag{3.12}$$

where

$$\Lambda^{1/r} = \text{diag}\{\lambda_1^{1/r}, \ldots, \lambda_n^{1/r}\}.$$

λ_i being a (positive) eigenvalue of A, and $\lambda_i^{1/r}$ the positive rth root of λ_i $(i = 1, 2, \ldots, n)$.

We now consider further theorems illustrating important properties of positive definite matrices.

Theorem 3.5
If $A \in \Sigma_n$, then there exists a non-singular matrix P such that

(1) $P^*AP = I$
(2) $PP^* = A^{-1}$

Proof
(1) Since $A \in \Sigma_n$, there exists a unitary matrix U such that

$$U^*AU = \Lambda$$

where

$$\Lambda = \text{diag}\{\lambda_1, \ldots, \lambda_n\},$$

λ_i being the eigenvalues of A.

Let

$$B = \Lambda^{-(1/2)} \quad \text{and} \quad P = UB. \tag{3.13}$$

Since U and B are invertible (remember $B^{-1} = \Lambda^{1/2}$),

$$(UB)^{-1} = B^{-1}U^{-1} = P^{-1}$$

hence P is non-singular.

Also

$$\begin{aligned} P^*AP &= B^*U^*AUB \\ &= B^*\Lambda B \\ &= B\Lambda B \qquad (\text{since } B^* = B = \Lambda^{-(1/2)}) \\ &= \Lambda B^2 \\ &= \Lambda(\Lambda^{-(1/2)})^2 = I. \end{aligned}$$

(2) Since by (1)

$$P^*AP = I$$

it follows that

$$P^*APP^* = P^*.$$

Since P is invertible, so is P^*. Hence

$$(P^*)^{-1}P^*APP^* = I$$

or

$$APP^* = I$$

The result follows.

Example 3.4

Given

$$A = \begin{bmatrix} 8 & -2i \\ 2i & 5 \end{bmatrix}$$

find the matrix P such that

$$P^*AP = I.$$

Solution

In Example 3.3 we found that

$$\Lambda^{-(1/2)} = \text{diag}\left\{\frac{1}{2}, \frac{1}{3}\right\} \quad \text{and} \quad U = \frac{1}{\sqrt{10}}\begin{bmatrix} 1+i & 2-2i \\ 2-2i & 1+i \end{bmatrix}$$

it follows by (3.13) that

$$P = \frac{1}{6\sqrt{10}}\begin{bmatrix} 3(1+i) & 4(1-i) \\ 6(1-i) & 2(1+i) \end{bmatrix}.$$

It is easy to check that

$$P^*AP = I$$

and that

$$PP^* = \frac{1}{36}\begin{bmatrix} 5 & 2i \\ -2i & 8 \end{bmatrix} = A^{-1}.$$

Theorem 3.6
$A \in \Sigma_n$ if and only if there exists a non-singular matrix R such that

$$A = R^*R$$

Proof
If $A \in \Sigma_n$ the result follows immediately from the previous theorem. Since $A^{-1} = PP^*$, where P is non-singular, it follows that $A = R^*R$, where $R = P^{-1}$.

Assume $A = R^*R$.

Then

$$A^* = (R^*R)^* = R^*R = A,$$

hence A is Hermitian.

If $\mathbf{x} \neq \mathbf{0}$, then $\mathbf{y} = R\mathbf{x} \neq \mathbf{0}$ since R is non-singular.

Also

$$(\mathbf{x}, A\mathbf{x}) = (\mathbf{x}, R^*R\mathbf{x}) = (\mathbf{x}, R^*\mathbf{y}) = (R\mathbf{x})^*\mathbf{y} = (\mathbf{y}, \mathbf{y}) > 0.$$

It follows that $A \in \Sigma_n$.

Example 3.5
Write the matrix

$$A = \begin{bmatrix} 8 & -2i \\ 2i & 5 \end{bmatrix}$$

as a product R^*R.

Solution
In Example 3.4 we found the matrix P.

$$R = P^{-1} = \frac{1}{\sqrt{10}}\begin{bmatrix} 2(1-i) & 4(1+i) \\ 6(1+i) & 3(1-i) \end{bmatrix}. \quad ■$$

We have considered various important properties of positive definite matrices, but it is also worthwhile to point out two fundamental properties *not* possessed by such matrices.

For example, if $A \in \Sigma_n$, then $(-A) \notin \Sigma_n$. The implication of this is that Σ_n is not an additive group.

If $A, B \in \Sigma_n$, then $AB \notin \Sigma_n$ (in general).
For example both

$$A = \begin{bmatrix} 1 & 2 \\ 2 & 5 \end{bmatrix} \quad \text{and} \quad B = \begin{bmatrix} 1 & -2 \\ -2 & 5 \end{bmatrix} \in \Sigma_2$$

but

$$AB = \begin{bmatrix} -3 & 8 \\ -8 & 21 \end{bmatrix} \notin \Sigma_2.$$

The implication of this is that Σ_n is not a multiplicative group.

In spite of the fact that Σ_n is not a group, it does possess various group properties as illustrated by the following theorem.

Theorem 3.7

Consider $A, B \in \Sigma_n$,

$c > 0$ a scalar, r a positive integer, m any integer and p a rational number.

Then

(1) $A + B \in \Sigma_n$
(2) $cA \in \Sigma_n$
(3) $A^m \in \Sigma_n$
(4) $A^{1/r} \in \Sigma_n$
(5) $A^p \in \Sigma_n$.

Proof

(1) $(A + B)^* = A^* + B^* = A + B$, hence $(A + B)$ is Hermitian.
If

$$\mathbf{x} \neq \mathbf{0}$$

$$(\mathbf{x}, [A + B]\mathbf{x}) = (\mathbf{x}, A\mathbf{x}) + (\mathbf{x}, B\mathbf{x}) > 0.$$

The result follows.

(2) $(cA)^* = cA^* = cA$, hence cA is Hermitian.
If

$$\mathbf{x} \neq \mathbf{0}$$

$$(\mathbf{x}, cA\mathbf{x}) = c(\mathbf{x}, A\mathbf{x}) > 0 \quad \text{since} \quad c > 0.$$

The result follows.

(3) $(A^m)^* = (A^*)^m = A^m$, hence A^m is Hermitian.
Assume

$$m > 0.$$

Since $A \in \Sigma_n$, there exists a unitary matrix U such that

$$A = U\Lambda U^*$$

where

$$\Lambda = \text{diag}\{\lambda_1, \ldots, \lambda_n\}, \ \lambda_i > 0 \ (\text{all } i).$$

Then

$$A^m = U\Lambda^m U^*,$$

so that

$$\begin{aligned} U^*A^m U &= \Lambda^m \\ &= \text{diag}\{\lambda_1^m, \ldots, \lambda_n^m\}, \ \ \lambda_i^m > 0. \end{aligned}$$

The result follows, by Theorem 3.2.
If

$$m = 0,$$

then

$$A^m = I \in \Sigma_n.$$

If $m = -1$, then by Theorem 3.5 there exists a non-singular matrix P such that

$$PP^* = A^{-1}.$$

Hence by Theorem 3.6,

$$A^{-1} \in \Sigma_n.$$

If

$$m < -1, \ \ A^m = (A^{-1})^{-m} = B^{-m}$$

where

$$B = A^{-1} \in \Sigma_n$$

by the last statement. Since

$$(-m) > 0, \ \ B^{-m} = A^m \in \Sigma_n.$$

(4) By (3.12)

$$A^{1/r} = U\Lambda^{1/r}U^*$$

where

$$\Lambda^{1/r} = \text{diag}\{\lambda_1^{1/r}, \ldots, \lambda_n^{1/r}\} \ \ (\lambda_i > 0)$$

where $\lambda_i^{1/r}$ is the positive rth root of λ_i.

Also

$$(A^{1/r})^* = (U\Lambda^{1/r}U^*)^* = U\Lambda^{1/r}U^* = A^{1/r}.$$

Hence $A^{1/r}$ is Hermitian. By Theorem (3.2), $A^{1/r} \in \Sigma_n$.

(5) $A^p \in \Sigma_n$ by (3) and (4) above.

PROBLEMS

3.1 Write the following quadratic forms in matrix notation.

(1) $3x^2 - 2y^2 - 8xy + xz + z^2$.

(2) $x^2 - 2y^2 + 3z^2 + 2xy - 4xz + 6yz$.

3.2 Obtain the unitary matrix U such that

$$U^*AU$$

is in a diagonal form when

(1) $A = \begin{bmatrix} 3 & 1 & 1 \\ 1 & 0 & 2 \\ 1 & 2 & 0 \end{bmatrix}$

(2) $A = \begin{bmatrix} 7 & -2 & 1 \\ -2 & 10 & -2 \\ 1 & -2 & 7 \end{bmatrix}$.

3.3 A is a real (symmetric) positive definite matrix of order $(n \times n)$. Show that

(1) A^{-1} exists and is positive definite.

(2) A^m is positive definite for all integers m.

3.4 $P(n \times n)$ is a positive definite matrix.
$S(n \times n)$ is a positive semi-definite matrix.
Show that the sum

$$P + S$$

is a positive definite matrix.

3.5 Investigate the nature of the curve given by

(1) $x_1^2 + 8x_1x_2 + x_2^2 = 1$

(2) $x_1^2 + 4x_1x_2 - 2x_2^2 = 5$

(3) $5x_1^2 - 2x_1x_2 + 5x_2^2 = 2$.

3.6 Examine for definiteness the matrix

(1) $A = \begin{bmatrix} 3 & -1 & 3 \\ -1 & 5 & 1 \\ 3 & 1 & 5 \end{bmatrix}$

(2) $A = \begin{bmatrix} -2 & 1 & -1 \\ 1 & -2 & 1 \\ -1 & 1 & -2 \end{bmatrix}$

(3) $A = \begin{bmatrix} 2 & 1 & -2 \\ 1 & 1 & -3 \\ -2 & -3 & 10 \end{bmatrix}$.

3.7 $A(n \times n)$ is a real matrix of rank r. Show that

(1) AA' is a positive definite if $r = n$

(2) AA' is positive semi-definite if $r < n$.

3.8 Consider a quadratic form

$$q(\mathbf{x}) = \mathbf{x}'A\mathbf{x}$$

where $\mathbf{x}$ is a unit vector.

Let $U = [\mathbf{u}_1, \mathbf{u}_2, \ldots, \mathbf{u}_n]$ be the modal matrix of A where $\mathbf{u}_i$ $(i = 1, 2, \ldots, n)$ is an orthonormal set of the eigenvectors of A. Show that

$$q(\mathbf{x}) = \sum_{i=1}^{n} \lambda_i \cos^2 \theta_i$$

where λ_i $(i = 1, 2, \ldots, n)$ are the eigenvalues of A and

$$\cos \theta_i = (\mathbf{x}_j, \mathbf{u}_i), \quad (i = 1, 2, \ldots, n).$$

(This result is known as the *Euler's* theorem.)

4

Nonnegative Matrices

4.1 INTRODUCTION

In various important applications of matrix theory, the elements of the matrices considered are all nonnegative. The usefulness of this theory to Linear Programming, Stochastic Processes, solving certain types of partial differential equations, various econometric models and other applications, has grown rapidly and this has in turn stimulated further research which is expanding at an impressive rate.

Much of the work on nonnegative matrices is based on their properties discovered by Perron and Frobenius and elegantly proved by Wielandt [27].

4.2 TERMINOLOGY AND NOTATION

Let $A = [a_{ij}]$ and $B = [b_{ij}]$ be matrices of order $(m \times n)$. We write

$$A \geqslant B \text{ if } a_{ij} \geqslant b_{ij} \qquad (\text{all } i, j)$$

$$A > B \text{ if } a_{ij} > b_{ij} \qquad (\text{all } i, j)$$

In particular, A is said to be *nonnegative* if

$$A \geqslant 0 \quad \text{and} \quad A \neq 0$$

and is *positive* if

$$A > 0.$$

The same notation is used for (row or column) vectors.

If

$$\mathbf{x} = [x_1 x_2 \ldots x_n]' \quad \text{and} \quad \mathbf{y} = [y_1 y_2 \ldots y_n]',$$

then

$$\mathbf{y} \geqslant \mathbf{x}$$

if

$$y_i \geqslant x_i \qquad \text{(all } i\text{)}.$$

In particular

$$\mathbf{y} > \mathbf{0} \quad \text{iff} \quad y_i > 0 \qquad \text{(all } i\text{)}.$$

In general when writing $\mathbf{y} \geqslant \mathbf{0}$, we exclude the possibility that $\mathbf{y} = \mathbf{0}$.

Example 4.1
Let

$$A(n \times n) > 0 \quad \text{and} \quad \mathbf{x} \geqslant \mathbf{0} \quad (\text{but } \mathbf{x} \neq \mathbf{0})$$

Show that

$$A\mathbf{x} = \mathbf{y} > \mathbf{0}.$$

Solution

$$\begin{aligned} y_i &= a_{i1}x_1 + a_{i2}x_2 + \ldots + a_{in}x_n \\ &\geqslant a_{ik}(x_1 + x_2 + \ldots + x_n) \end{aligned}$$

where

$$a_{ik} = \min\{a_{i1}, a_{i2}, \ldots, a_{in}\} > 0 \quad (\text{as } A > 0).$$

Since

$$\mathbf{x} \geqslant \mathbf{0}$$

then

$$\sum_i x_i > 0,$$

hence

$$y_i > 0 \qquad (i = 1, 2, \ldots, n).$$

The result follows. ■

Although the notation for vectors and matrices has been explained in Section 1.1, a word of warning is necessary which is due to the author's anxiety to keep the notation as simple as possible.

We use vector notation $\mathbf{x}$ in the usual manner. The problem arises when a set of vectors is considered, say

$$\mathbf{x}_i, \mathbf{x}_2, \ldots, \mathbf{x}_r$$

where each has n components (say)

$$\mathbf{x}_i = [x_1 x_2 \ldots x_n]'.$$

So rather than use a double suffix notation, the reader is asked to be careful to discriminate between vectors and scalars. Of course when the distinction is not clear, the double suffix notation is used.

4.3 THE PERRON–FROBENIUS THEOREM FOR A POSITIVE MATRIX

To prove the Perron–Frobenius theorem for positive square matrices, we need to establish a number of inequalities.

We consider a positive matrix

$$A(n \times n) = [a_{ij}]$$

where

$$A > 0$$

and evaluate

$$\phi(\mathbf{x}) = \min_j \left\{ \frac{\Sigma\, a_{jr} x_r}{x_j} \right\} \qquad (j = 1, 2, \ldots, n) \tag{4.1}$$

for vector

$$\mathbf{x} = [x_1 x_2 \ldots x_n]' \geqslant \mathbf{0}.$$

Values of $\phi(\mathbf{x})$ for which $x_j = 0$ are excluded when evaluating the minimum.

Example 4.2
For

$$A = \begin{bmatrix} 2 & 2 & 1 \\ 1 & 3 & 1 \\ 1 & 2 & 2 \end{bmatrix}$$

and for each of the vectors

$$\mathbf{x}_1 = [1 \;\; 2 \;\; 1]' \qquad \mathbf{x}_4 = \begin{bmatrix} \frac{1}{\sqrt{6}} & \frac{2}{\sqrt{6}} & \frac{1}{\sqrt{6}} \end{bmatrix}'$$

$$\mathbf{x}_2 = [1 \;\; 1 \;\; 0]' \qquad \mathbf{x}_5 = \begin{bmatrix} \frac{1}{\sqrt{2}} & \frac{1}{\sqrt{2}} & 0 \end{bmatrix}'$$

$$\mathbf{x}_3 = [1 \;\; 1 \;\; 1]' \qquad \mathbf{x}_6 = \begin{bmatrix} \frac{1}{\sqrt{3}} & \frac{1}{\sqrt{3}} & \frac{1}{\sqrt{3}} \end{bmatrix}'$$

evaluate

$$\mathbf{w}_i = A\mathbf{x}_i \qquad (i = 1, 2, \ldots, 6)$$

and then

$$\phi(\mathbf{x}) = \min_j \left\{ \frac{w_j}{x_j} \right\} \qquad (x_j \neq 0)$$

where

$$w_j = \sum_r a_{jr} x_r \qquad (j = 1, 2, 3).$$

Also evaluate

$$\mathbf{z} = \phi(\mathbf{x})\mathbf{x}$$

Solution
Let

$$A\mathbf{x}_i = \mathbf{w}_i \qquad (i = 1, 2, \ldots, 6)$$

and then

$$\mathbf{z}_i = \phi(\mathbf{x}_i)\mathbf{x}_i$$

$$\mathbf{w}_1 = [7 \;\; 8 \;\; 7]' \qquad \mathbf{w}_4 = \left[\frac{7}{\sqrt{6}} \;\; \frac{8}{\sqrt{6}} \;\; \frac{7}{\sqrt{6}}\right]'$$

$$\mathbf{w}_2 = [4 \;\; 4 \;\; 3]' \qquad \mathbf{w}_5 = \left[\frac{1}{\sqrt{2}} \;\; \frac{4}{\sqrt{2}} \;\; \frac{3}{\sqrt{2}}\right]'$$

$$\mathbf{w}_3 = [5 \;\; 5 \;\; 5]' \qquad \mathbf{w}_6 = \left[\frac{5}{\sqrt{3}} \;\; \frac{5}{\sqrt{3}} \;\; \frac{5}{\sqrt{3}}\right]'$$

$$\phi(\mathbf{x}_1) = \min\left\{\frac{7}{1}, \frac{8}{2}, \frac{7}{1}\right\} = 4 \qquad \phi(\mathbf{x}_4) = \min\left\{\frac{7}{1}, \frac{8}{2}, \frac{7}{1}\right\} = 4$$

$$\phi(\mathbf{x}_2) = \min\left\{\frac{4}{1}, \frac{4}{1}\right\} = 4 \qquad \phi(\mathbf{x}_5) = \min\left\{\frac{4}{1}, \frac{4}{1}\right\} = 4$$

$$\phi(\mathbf{x}_3) = \min\left\{\frac{5}{1}, \frac{5}{1}, \frac{5}{1}\right\} = 5 \qquad \phi(\mathbf{x}_6) = \min\left\{\frac{5}{1}, \frac{5}{1}, \frac{5}{1}\right\} = 5$$

$$\mathbf{z}_1 = [4 \;\; 8 \;\; 4]' \qquad \mathbf{z}_4 = \left[\frac{4}{\sqrt{6}} \;\; \frac{8}{\sqrt{6}} \;\; \frac{4}{\sqrt{6}}\right]'$$

$$\mathbf{z}_2 = [4\ \ 4\ \ 0]' \qquad \mathbf{z}_5 = \begin{bmatrix} \frac{4}{\sqrt{2}} & \frac{4}{\sqrt{2}} & 0 \end{bmatrix}'$$

$$\mathbf{z}_3 = [5\ \ 5\ \ 5]' \qquad \mathbf{z}_6 = \begin{bmatrix} \frac{5}{\sqrt{3}} & \frac{5}{\sqrt{3}} & \frac{5}{\sqrt{3}} \end{bmatrix}'.$$

Notes

1. In the way $\phi(\mathbf{x})$ has been defined in (4.1), we see that for a given $\mathbf{x}$, $\phi(\mathbf{x})$ is the largest number υ (say) such that

$$\mathbf{w} \geqslant \mathbf{z},$$

that is

$$A\mathbf{x} \geqslant \upsilon\mathbf{x}. \tag{4.2}$$

To appreciate this important result, note that for $\mathbf{x} = [1\ 2\ 1]'$, $\mathbf{w} = [7\ 8\ 7]'$ we found above that $\phi(\mathbf{x}) = \upsilon = 4$, and $\mathbf{z} = [4\ 8\ 2]'$, so that (4.2) is satisfied.

But, for argument sake, take a different (larger) value for υ, say $\upsilon^* = 4.1$, then $z = [4.1\ 8.2\ 4.1]'$, and this time (4.2) is *not* satisfied.

(2) From the definition (4.1) it is clear that

$$\phi(\mathbf{x}) \quad \text{and} \quad \phi(k\mathbf{x})$$

for any scalar $k > 0$, have the same value. This accounts for $\phi(\mathbf{x}_1)$ and $\phi(\mathbf{x}_4)$, $\phi(\mathbf{x}_2)$ and $\phi(\mathbf{x}_5)$, $\phi(\mathbf{x}_3)$ and $\phi(\mathbf{x}_6)$ having the same values.

So when investigating the values of $\phi(\mathbf{x})$ for a matrix A and for all $\mathbf{x} \geqslant \mathbf{0}$ we need only consider, without loss of generality, the values of $\phi(\mathbf{x})$ for normalised $\mathbf{x} \geqslant 0$, that is for

$$\left\{\mathbf{x} : \sum_i x_i^2 = 1\right\}.$$

3. In the definition (4.1) we have omitted the consideration of $\phi(\mathbf{x})$ for all $x_j = 0$. The zero elements of $\mathbf{x}$ obviously complicate the analysis of the function involved. Fortunately such situations can be avoided, again without loss of generality.

Since by (4.2)

$$A\mathbf{x} \geqslant \phi(\mathbf{x})\mathbf{x}$$

on multiplying both sides by A

$$A(A\mathbf{x}) \geqslant \phi(\mathbf{x})A\mathbf{x}$$

or

$$A\mathbf{y} \geqslant \phi(\mathbf{x})\mathbf{y} \tag{4.3}$$

where

$$\mathbf{y} = A\mathbf{x} > \mathbf{0} \qquad \text{(by Equation 4.1)}$$

By an argument analogous to the above

$$A\mathbf{y} \geqslant \phi(\mathbf{y})\mathbf{y} \tag{4.4}$$

where $\phi(\mathbf{y})$ is the *largest* number for which the inequality holds. Hence by (4.3) and (4.4), for all $\mathbf{x} \geqslant \mathbf{0}$

$$\phi(\mathbf{y})\mathbf{y} \geqslant \phi(\mathbf{x})\mathbf{y}$$

or

$$\phi(\mathbf{y}) \geqslant \phi(\mathbf{x}) \tag{4.5}$$

where

$$\mathbf{y} \in S = \left\{\mathbf{y} : \mathbf{y} = A\mathbf{x} > \mathbf{0};\ \sum_i x_i^2 = 1\right\}.$$

Note

It will be appreciated that vectors in S are a subset of the set of non-negative vectors $\{\mathbf{x}\}$.

The set S defined above is closed and bounded, since it is a continuous mapping $\mathbf{y} = A\mathbf{x}$ of $\{\mathbf{x}\}$ for which $\mathbf{x} \geqslant \mathbf{0}$ and $\Sigma x_i^2 = 1$. Therefore on this set, $\phi(\mathbf{x})$ attains its maximum r, hence

$$r = \max_{\mathbf{x} \geqslant \mathbf{0}} \{\phi(\mathbf{x})\} = \max_{\mathbf{y} \in S} \{\phi(\mathbf{y})\} \tag{4.6}$$

This maximum r is attained for some $\mathbf{y}_0 \in S$.

Hence

$$r = \max \phi(\mathbf{x}) = \max \phi(\mathbf{y}) = \phi(\mathbf{y}_0) \tag{4.7}$$

We shall now prove that r is an eigenvalue of A.

By (4.2)

$$A\mathbf{y}_0 \geqslant r\mathbf{y}_0$$

Assume that

$$A\mathbf{y}_0 \neq r\mathbf{y}_0$$

so that

$$[A - rI]\,\mathbf{y}_0 \geqslant \mathbf{0}$$

(at least one element of this vector is $\neq 0$).

Then

$$[A - rI]\,\mathbf{u} > \mathbf{0}$$

where

$$\mathbf{u} = A\mathbf{y}_0$$

or

$$A\mathbf{u} > r\mathbf{u}.$$

But by definition, $\phi(\mathbf{u})$ is the *largest* number for which the inequality

$$A\mathbf{u} \geqslant \phi(\mathbf{u})\mathbf{u} \text{ holds.} \tag{4.8}$$

We conclude that

$$\phi(\mathbf{u}) > r$$

which contradicts the definition of r.

It follows that

$$A\mathbf{y}_0 = r\mathbf{y}_0 \tag{4.9}$$

(4.9) shows that r is an eigenvalue of the matrix A and $\mathbf{y}_0$ is a corresponding eigenvector. Notice that $\mathbf{y}_0 > \mathbf{0}$ (since $\mathbf{y}_0 \in S$). From the above discussion we deduce the following important results which forms part of the famous Perron–Frobenius theorem for positive matrices.

Theorem 4.1

(1) The eigenvalue r of $A > 0$ defined by (4.6) is positive.
(2) r is the spectral radius (ρ) of A (see Section 1.7).
(3) The eigenvector $\mathbf{y}_0$ in (4.9) corresponding to the eigenvalue r is positive.
(4) The multiplicity of the eigenvalue r of A is one.

Proof

(1) Let

$$\mathbf{x} = \left[\frac{1}{\sqrt{n}} \;\; \frac{1}{\sqrt{n}} \;\cdots\; \frac{1}{\sqrt{n}}\right]',$$

then by (4.1)

$$\phi(\mathbf{x}) = \min_j \left\{\sum_r a_{jr}\right\} > 0 \quad \text{as} \quad A > 0$$

Since $r \geqslant \phi(\mathbf{x})$ we conclude that $r > 0$.

(2) Assume that λ is an eigenvalue of A and $\mathbf{x}$ the corresponding eigenvector, (both λ and $\mathbf{x}$ may be complex). Then

$$A\mathbf{x} = \lambda\mathbf{x}$$

that is

$$\sum_{r=1}^{n} a_{jr}x_r = \lambda x_j \qquad (j = 1, 2, \ldots, n).$$

On taking the moduli of both sides, we obtain

$$|\lambda|\,|x_j| = \left|\sum_{r=1}^{n} a_{jr}\,x_r\right| \leqslant \sum_{r=1}^{n} a_{jr}\,|x_r| \qquad (j = 1, 2, \ldots, n).$$

Let $\tilde{\mathbf{x}} = [|x_1|\,|\,x_2|\ldots|x_n|]'$, then from the above we obtain

$$|\lambda|\,\tilde{\mathbf{x}} \leqslant A\tilde{\mathbf{x}}$$

But by (4.2), $\phi(\tilde{\mathbf{x}})$ is the *largest* number such that

$$A\tilde{\mathbf{x}} \geqslant \phi(\tilde{\mathbf{x}})\tilde{\mathbf{x}}.$$

Hence

$$\phi(\tilde{\mathbf{x}}) \geqslant |\lambda|$$

and by (4.6)

$$r \geqslant \phi(\tilde{\mathbf{x}}) \geqslant |\lambda| \tag{4.10}$$

so (2) follows.

(3) This follows immediately since $\mathbf{y}_0 \in S$ and every vector in S is positive.

(4) This part of the proof will be considered in Theorem 4.3. ■

We can also prove part (3) above in the following manner.

Assume that

$$|\lambda| = r$$

in (4.10) above, then

$$r = \phi(\tilde{\mathbf{x}}) = |\lambda|.$$

It follows from the definitions of $\phi(\mathbf{x})$ and r that

$$A\tilde{\mathbf{x}} = r\tilde{\mathbf{x}}$$

showing that

$$\tilde{\mathbf{x}} = [|x_1|,\ |x_2|, \ldots, |x_n|]'$$

is an eigenvector of A corresponding to the eigenvalue r.

Also

$$r\,|x_i| = \sum_{j=1}^{n} a_{ij}\,|x_j| \qquad (i = 1, 2, \ldots, n).$$

Since all $a_{ij} > 0$ and not all $|x_j|$ can be zero (since $\tilde{\mathbf{x}}$ is an eigenvector), it follows that

$$r|x_i| > 0,$$

so that

$$|x_i| > 0 \qquad (i = 1, 2, \ldots, n).$$

Example 4.3

Consider a matrix $A > 0$ having a spectral radius $r = \rho(A)$.

Assume that A has another eigenvalue λ such that

$$|\lambda| = r.$$

Show that the corresponding eigenvector $\mathbf{x}$ has components

$$\mathbf{x} = [x_1, x_2, \ldots, x_n]'$$

which are such that

$$x_j = |x_j| e^{i\theta} \qquad (j = 1, 2, \ldots, n)$$

and hence that the above assumption is false so that

$$r > |\lambda|$$

for all other eigenvalues λ of A.

Solution

$$\lambda \mathbf{x} = A\mathbf{x},$$

so that

$$|\lambda|\,|x_j| = |\sum a_{js} x_s| \leqslant \sum a_{js}\,|x_s| \qquad (j = 1, 2, \ldots, n).$$

It has been shown above that if

$$|\lambda| = r,$$

then

$$r\tilde{\mathbf{x}} = A\tilde{\mathbf{x}},$$

so that,

$$r|x_j| = \sum a_{js}|x_s| \qquad (j = 1, 2, \ldots, n)$$

It follows that (when $|\lambda| = r$)

$$|\sum a_{js} x_s| = \sum a_{js}\,|x_s|.$$

From elementary theory of complex numbers we know that

$$|z_1 + z_2| = |z_1| + |z_2|$$

if

$$z_1\bar{z}_2 + \bar{z}_1 z_2 = 2|z_1|\,|z_2|.$$

Writing

$$z_1 = |z_1|e^{i\theta_1} \quad \text{and} \quad z_2 = |z_2|e^{i\theta_2}$$

it follows that

$$\theta_1 = \theta_2 = \theta \quad \text{(say)}.$$

Since $a_{ij} > 0$ (all i, j), the generalisation of this result leads to the solution

$$\mathbf{x} = \tilde{\mathbf{x}}e^{i\theta}$$

showing that $\mathbf{x}$, being a constant multiple of $\tilde{\mathbf{x}}$, is an eigenvector corresponding to the eigenvalue r of A.

It follows that

$$\lambda = r.$$

So there can be no other eigenvalue λ of A (except for $\lambda = r$) such that $|\lambda| = r$.

Since $r = \rho(A)$ it follows that

$$r > |\lambda|$$

for all other eigenvalue λ of A.

Theorem 4.2

Consider matrices A and B such that

$$A \geqslant B > 0.$$

(1) If r is the spectral radius of A and λ is an eigenvalue of B, then

$$|\lambda| \leqslant r$$

(2) If $|\lambda| = r$, then $A = B$.

Proof

(1) Since the eigenvalues of A are also the eigenvalues of A', we have

$$A'\mathbf{x} = r\mathbf{x} \tag{4.11}$$

where $\mathbf{x} > \mathbf{0}$ is the eigenvector of A' corresponding to the eigenvalue r. Also

$$B\mathbf{y} = \lambda\mathbf{y}$$

where $\mathbf{y} = [y_1 y_2 \ldots y_n]'$ is an eigenvector of B corresponding to the eigenvalue λ.

Since

$$|\lambda|\tilde{\mathbf{y}} \leqslant B\tilde{\mathbf{y}}$$

it follows that

$$|\lambda|\tilde{\mathbf{y}} \leqslant B\tilde{\mathbf{y}} \leqslant A\tilde{\mathbf{y}} \tag{4.12}$$

where $\tilde{\mathbf{y}} = [|y_1|\,|y_2|\ldots|y_n|]'$.

Premultiplying (4.12) by $\mathbf{x}'$, we obtain

$$|\lambda|\,\mathbf{x}'\tilde{\mathbf{y}} \leqslant \mathbf{x}'B\tilde{\mathbf{y}} \leqslant \mathbf{x}'A\tilde{\mathbf{y}} = r\mathbf{x}'\tilde{\mathbf{y}} \qquad \text{(by (4.11))}.$$

But $\mathbf{x}'\tilde{\mathbf{y}} > 0$ (since $\mathbf{x} > \mathbf{0}$ and $\tilde{\mathbf{y}} \geqslant \mathbf{0}$) hence

$$|\lambda| \leqslant r.$$

(2) If $|\lambda| = r$ then by (4.12)

$$r\tilde{\mathbf{y}} \leqslant A\tilde{\mathbf{y}}$$

so that

$$\mathbf{z} = A\tilde{\mathbf{y}} - r\tilde{\mathbf{y}} \geqslant \mathbf{0}.$$

Hence

$$\mathbf{z} \neq \mathbf{0} \quad \text{or} \quad \mathbf{z} = \mathbf{0}.$$

Assume that $\mathbf{z} \neq \mathbf{0}$.

Then (see Example 4.1) $A\mathbf{z} > \mathbf{0}$, so that

$$A\mathbf{w} > r\mathbf{w}$$

where

$$A\tilde{\mathbf{y}} = \mathbf{w} = [w_1 w_2 \ldots w_n]' \qquad \text{(say)}$$

It follows that

$$\frac{\sum a_{js} w_s}{w_j} > r \qquad (j = 1, 2, \ldots, n).$$

This contradicts the definition of r (see (4.6) and (4.1)). We conclude that

$$\mathbf{z} = \mathbf{0}.$$

Hence from (4.12)

$$r\tilde{\mathbf{y}} = B\tilde{\mathbf{y}} = A\tilde{\mathbf{y}}$$

where

$$\tilde{\mathbf{y}} > \mathbf{0}.$$

As $A \geqslant B$, it follows that $A = B$.

Note

Although it has been found convenient to prove this theorem for a positive matrix A, it is true for any nonnegative irreducible matrix A. The applications of the above result, is particularly useful when we consider a matrix $A > 0$ and the (partitioned) matrix

$$B = \left[\begin{array}{c:c} A(\alpha) & 0 \\ \hdashline 0 & 0 \end{array}\right] \tag{4.13}$$

where (as an example) $A(\alpha)$ is a principal submatrix of A. Then $A \geqslant B$, and the theorem (taking into account the above note) states that every eigenvalue of B is in modulus smaller (or equal) to the spectral radius r of A. The important fact here is that if λ is an eigenvalue of $A(\alpha)$, it is also an eigenvalue of B, hence $r \geqslant |\lambda|$. Notice the similarity of this result and the interlacing property for symmetric matrices (see Theorem 2.3).

Example 4.3

Illustrate the result of the above theorem for the matrix

$$A = \begin{bmatrix} 2 & 2 & 1 \\ 1 & 3 & 1 \\ 1 & 2 & 2 \end{bmatrix}$$

and all its principal submatrices.

Solution

The characteristic equation for A is

$$C(\lambda) = \lambda^3 - 7\lambda^2 + 11\lambda - 5 = 0,$$

so that

$$r = \lambda_1 = 5 \quad \text{and} \quad \lambda_2 = \lambda_3 = 1.$$

For

$$\alpha = \{2,\ 3\} \qquad A(\alpha) = \begin{bmatrix} 3 & 1 \\ 2 & 2 \end{bmatrix}, \quad C(\lambda) = (\lambda - 1)(\lambda - 4)$$

for

$$\alpha = \{1,\ 2\} \qquad A(\alpha) = \begin{bmatrix} 2 & 2 \\ 1 & 3 \end{bmatrix}, \quad C(\lambda) = (\lambda - 1)(\lambda - 4)$$

for

$$\alpha = \{1,\ 3\} \qquad A(\alpha) = \begin{bmatrix} 2 & 1 \\ 1 & 2 \end{bmatrix}, \quad C(\lambda) = (\lambda - 1)(\lambda - 3).$$

In each case $r = 5$ is larger than the eigenvalues of $A(\alpha)$. ■

Theorem 4.3 which follows is a part of the Perron–Frobenius theorem for positive matrices.

Theorem 4.3
Given that $A > 0$ has a spectral radius $\rho(A) = r$, then $\lambda = r$ is a simple eigenvalue of A.

Proof
Let

$$c(s) = |sI - A|.$$

We have the following results from Section 1.8

(1) $c'(s)$ is a linear combination of the principal minors of

$$g(s) = [sI - A].$$

(2) If $c'(\lambda) \neq 0$, then the eigenvalue λ of A is simple.

A typical principal minor of $g(s)$ is

$$|sI - A(\alpha)|$$

where $A(\alpha)$ is a principal submatrix of A.

Since

$$|sI - A(\alpha)| = \prod(s - \lambda_i)$$

and by Theorem 4.2,

$$r = \rho(A) > |\lambda_i|;$$

it follows that

$$|sI - A(\alpha)| \neq 0 \quad \text{for} \quad s \geqslant r.$$

In particular

$$|rI - A(\alpha)| > 0$$

and by (1) above

$$c'(r) > 0.$$

The result follows by (2) above.

Another interesting and important consequence of Theorem 4.2 is the following:

Theorem 4.4
If $B > 0$ has a spectral radius r, then if any element of B is increased, the spectral radius also increases.

Proof
Let A be the matrix obtained from B when the element of B is increased. Then

$$A \geqslant B$$

The result now follows by Theorem 4.2

Note
As for Theorem 4.2, this result applies when B is a nonnegative irreducible matrix.

Example 4.5
Illustrate the result of Theorem 4.4 when

$$B = \begin{bmatrix} 2 & 2 & 1 \\ 1 & 3 & 1 \\ 1 & 2 & 2 \end{bmatrix} \quad \text{and} \quad A = \begin{bmatrix} 2 & 2 & 1 \\ 1 & 3 & 1 \\ 1 & 2 & 3 \end{bmatrix}$$

Solution
For B:

$$c(\lambda) = \lambda^3 - 7\lambda^2 + 11\lambda - 5 = 0$$

so that

$$r = \lambda_1 = 5, \quad \lambda_2 = \lambda_3 = 1.$$

For A:

$$c(\lambda) = \lambda^3 - 8\lambda^2 + 16\lambda - 9 = 0$$

so that

$$r = \lambda_1 = 5.303, \quad \lambda_2 = 1.697 \quad \text{and} \quad \lambda_3 = 1.$$

Hence

$$\rho(A) > \rho(B).$$

The spectral radius of a positive matrix A is seen to be at the kernel of much of the discussion of its properties. It is therefore useful to have an estimate of the spectral radius without carrying out the very tedious task of setting up and solving the characteristic equations of A. The following theorem provides such an estimate.

Theorem 4.5
Assume that $A = [a_{ij}] > 0$ has spectral radius r. Then

(1) $$\min_i \sum_{j=1}^{n} a_{ij} \leqslant r \leqslant \max_i \sum_{j=1}^{n} a_{ij} \tag{4.14}$$

(2) If any row sum equals r in (4.14), all row sums are equal (and similarly for the maximal and minimal column sums).

Note
Again, these results can be generalised and shown to apply to irreducible matrices.

Proof

(1) By (4.6), for any $\mathbf{x} \geqslant \mathbf{0}$

$$r \geqslant \phi(\mathbf{x}) = \min_j \left\{ \frac{\sum_{i=1}^{n} a_{ji} x_i}{x_j} \right\} \qquad j = 1, 2, \ldots, n$$

We define the *one vector* as

$$\mathbf{e} = [1 \; 1 \; \ldots \; 1]'.$$

For $\mathbf{x} = \mathbf{e}$, the above inequality becomes

$$r \geqslant \phi(\mathbf{e}) = \min_j \left\{ \sum_{j=1}^{n} a_{ji} \right\} \tag{4.15}$$

By (4.2), for $\mathbf{x} \geqslant \mathbf{0}$

$$\mathbf{x}\,\phi(\mathbf{x}) \leqslant A\mathbf{x}$$

so that

$$\mathbf{e}'\mathbf{x}\phi(\mathbf{x}) \leqslant \mathbf{e}'A\mathbf{x}$$

and since

$$\mathbf{e}'\mathbf{x} > 0,$$

$$\phi(\mathbf{x}) \leqslant \frac{\mathbf{e}'A\mathbf{x}}{\mathbf{e}'\mathbf{x}}. \tag{4.16}$$

Let α_i = sum of the elements of the ith column of A

$$= \sum_{j=1}^{n} a_{ji} \qquad (i = 1, 2, \ldots, n)$$

and

$$\alpha = \max\{\alpha_1, \alpha_2, \ldots, \alpha_n\} = \max_i \left\{ \sum_{j=1}^{n} a_{ji} \right\}.$$

Then

$$\mathbf{e}'A = [\alpha_1, \alpha_2, \ldots, \alpha_n] \leqslant [\alpha, \alpha, \ldots, \alpha] = \alpha\mathbf{e}'$$

It follows that (4.16) can be written as

$$\phi(\mathbf{x}) \leqslant \frac{\mathbf{e}'A\mathbf{x}}{\mathbf{e}'\mathbf{x}} \leqslant \frac{\alpha\mathbf{e}'\mathbf{x}}{\mathbf{e}'\mathbf{x}} = \alpha = \max_i \left\{ \sum_{j=1}^{n} a_{ji} \right\} \tag{4.17}$$

Hence by (4.15) and (4.17)

$$\min_j \left\{ \sum_{i=1}^{n} a_{ji} \right\} \leqslant r \leqslant \max_i \left\{ \sum_{j=1}^{n} a_{ji} \right\} \tag{4.18}$$

Since A and A' have the same eigenvalues, on repeating the above argument for A', we obtain

$$\min_j \left\{ \sum_{i=1}^{n} a_{ij} \right\} \leqslant r \leqslant \max_i \left\{ \sum_{j=1}^{n} a_{ij} \right\} \tag{4.19}$$

(4.14) now follows on comparing (4.18) and (4.19).

(2) To simplify the argument, we consider the situation when all row sums of A equal r except for one which is smaller than r.

By increasing one element so that this row sum now also equals r, (4.14) still holds (with equality on both sides), but by Theorem 4.4, the spectral radius of the matrix is now greater than r.

Hence we have a contradiction between Theorem (4.4) and (4.14), so our assumption that one row is smaller than r must be false. We conclude that *all* row sums must be r.

Similar arguments apply if we assume that (1) more than one row sum is smaller than r, and (2) some row sums are greater and some smaller than r.

Example 4.6

Apply the results of Theorem 4.5 to

$$A = \begin{bmatrix} 2 & 2 & 1 \\ 1 & 3 & 1 \\ 1 & 2 & 3 \end{bmatrix} \quad \text{and} \quad B = \begin{bmatrix} 2 & 2 & 1 \\ 1 & 3 & 1 \\ 1 & 2 & 2 \end{bmatrix}$$

Solution

For matrix A, the row sums suggest that

$$5 \leqslant r \leqslant 6$$

(as do the column sums).

We know (see Example 4.5) that $r = 5.3$.

For matrix B, all row sums are equal to 5 hence $r = 5$. No column sum equals $r = 5$ (if one did equal $r = 5$, all would equal 5). In this case the column sums indicate that

$$4 \leqslant r \leqslant 7.$$

4.4 IRREDUCIBLE MATRICES

Having discussed various important properties of positive matrices, an immediate question arises. Do all or some of these properties apply to other nonnegative matrices?

Consider the following simple examples:

$$A_1 = \begin{bmatrix} 0 & 1 & 1 \\ 1 & 1 & 1 \\ 1 & 0 & 0 \end{bmatrix} \quad A_2 = \begin{bmatrix} 0 & 2 & 0 \\ 1 & 0 & 1 \\ 0 & 1 & 0 \end{bmatrix} \quad \text{and} \quad A_3 = \begin{bmatrix} 0 & 1 & 0 \\ 1 & 0 & 0 \\ 0 & 0 & 1 \end{bmatrix}$$

The eigenvalues and the eigenvector corresponding to the largest positive eigenvalue are as follows.

For

$$A_1 : \lambda_i = 2, 0, -1; \qquad \mathbf{x}' = [1 \;\; 2 \;\; 3]$$

For

$$A_2 : \lambda_i = \sqrt{3}' \, 0, -\sqrt{3}; \quad \mathbf{x}' = [2 \;\; \sqrt{3} \;\; 2]$$

For

$$A_3 : \lambda_i = 1, 1, -1; \qquad \mathbf{x}' = [1 \;\; 1 \;\; 0].$$

In each of these examples the spectral radius is a psotive eigenvalue $\lambda = r$ of the matrix.

In the first two examples, the eigenvector corresponding to r is positive.

In the third example, the eigenvector is nonnegative.

Again in the first two examples $\lambda = r$ is a simple eigenvalue. So it seems that there are matrices $A \geqslant 0$ which have most of the Perron–Frobenius properties we have proved for positive matrices.

But there are other matrices $A \geqslant 0$ which have only some of these properties. It is not difficult to show (see Section 1.13) that A_1 and A_2 above are irreducible matrices, whereas A_3 is (obviously) reducible.

This suggests that it may be desirable to investigate irreducible matrices and consider which of the above properties apply.

Theorem 4.6

If $A(n \times n)$ is a nonnegative irreducible matrix, then

$$[I + A]^{n-1}$$

is a positive matrix.

Proof

Let $\mathbf{y} \geqslant \mathbf{0}$ (see Section 4.2 for notation) and

$$\mathbf{z} = [I + A]\,\mathbf{y} = \mathbf{y} + A\mathbf{y}. \qquad (4.20) \qquad (4.20)$$

Since $A\mathbf{y} \geqslant \mathbf{0}$, (4.20) shows that $\mathbf{z}$ has *at least* as many non-zero elements as $\mathbf{y}$.

Assume that $\mathbf{z}$ and $\mathbf{y}$ have an equal number of non-zero elements.

By a rearrangement of the elements of $\mathbf{y}$ and $\mathbf{z}$ (if necessary), we can write

$$\mathbf{y} = \begin{bmatrix} \mathbf{u} \\ \cdots \\ \mathbf{0} \end{bmatrix} \quad \text{and } \mathbf{z} = \begin{bmatrix} \mathbf{v} \\ \cdots \\ \mathbf{0} \end{bmatrix}$$

where $\mathbf{u} > \mathbf{0}$ and $\mathbf{v} > \mathbf{0}$ have the same number of elements.

Writing A in a suitably partitioned form, (4.20) can be written as

$$\begin{bmatrix} \mathbf{v} \\ \hdashline \mathbf{0} \end{bmatrix} = \begin{bmatrix} \mathbf{u} \\ \hdashline \mathbf{0} \end{bmatrix} + \begin{bmatrix} A_{11} & A_{12} \\ \hdashline A_{21} & A_{22} \end{bmatrix} \begin{bmatrix} \mathbf{u} \\ \hdashline \mathbf{0} \end{bmatrix}$$

so that

$$\mathbf{0} = A_{21}\mathbf{u} \quad \text{or} \quad A_{21} = 0.$$

But this implies that A is reducible (see Section 1.12), so that the above assumption is false. We conclude that $\mathbf{z}$ has more non-zero elements than $\mathbf{y}$. Repeating this argument (at most) $(n-1)$ times, we conclude that

$$[I + A]^{n-1}\mathbf{y} > \mathbf{0}. \tag{4.21}$$

But $\mathbf{y} \geqslant \mathbf{0}$ is arbitrary, choose

$$\mathbf{y} = \mathbf{e}_i \qquad (i = 1, 2, \ldots, n)$$

where $\mathbf{e}_i$ is the ith unit vector (see Section 1.2) then (4.21) implies that all elements of each column of

$$[I + A]^{n-1}$$

are positive.

The result follows. ■

When considering the Perron–Frobenius theorem for irreducible matrices we will need a number of preliminary results, to begin with, some generalisations of the ones to be found in Section 1.7, which can be proved without much difficulty.

Consider a matrix $A(n \times n) \geqslant 0$ having eigenvalues $\lambda_1, \lambda_2, \ldots, \lambda_n$.

(1) The eigenvalues of

$$[I + A]^m$$

are

$$(1 + \lambda_1)^m, (1 + \lambda_2)^m, \ldots, (1 + \lambda_n)^m.$$

The converse is true for each real eigenvalue λ_i.

(2) The multiplicity of $\lambda = \lambda_i$ in A is k then the multiplicity of the eigenvalue $(1 + \lambda_i)^m$ in $[I + A]^m$ is at least k.

(3) If $\mathbf{x}$ is an eigenvector corresponding to the eigenvalue $(1 + \lambda)$ of $[I + A]$, so that

$$[I + A]\mathbf{x} = (1 + \lambda)\mathbf{x}$$

then

$$Ax = \lambda x,$$

so that $\mathbf{x}$ is an eigenvector corresponding to the eigenvalue λ of A.

(4) Next we must reconsider whether a maximum value of $\phi(\mathbf{x})$ defined in (4.1) exists for irreducible matrices. We proceed as in Section 4.2, and consider

$$S = \left\{\mathbf{y} : \mathbf{y} = A\mathbf{x}; \sum x_i^2 = 1\right\}$$

and

$$\phi(\mathbf{y}) = \min_j \left\{\frac{\sum a_{jr} y_r}{y_j}\right\}.$$

Whereas for a positive matrix A, $y_j > 0$ (all j), for an irreducible matrix it is possible that

$$y_j = 0 \quad \text{(for some } j\text{)}.$$

So $\phi(\mathbf{y})$ is not necessarily continuous on S.

We can overcome this difficulty by defining the set

$$Q = \left\{\mathbf{z} : \mathbf{z} = [I + A]^{n-1}\mathbf{y};\ \mathbf{y} \in S\right\}$$

So that $S \supset Q$.

Where $\mathbf{z} \in Q$, $z_i > 0$ (for all i), so that $\phi(\mathbf{z})$ is continuous on the closed and bounded set Q.

It follows that $\phi(\mathbf{z})$ attains a maximum value for some $\mathbf{z}_0 \in Q$, $\phi(\mathbf{z}_0) = r$ (say).

For each $\mathbf{y} \in S$,

$$A\mathbf{y} \geqslant \phi(\mathbf{y}) \cdot \mathbf{y}$$

so that

$$A\mathbf{z} \geqslant \phi(\mathbf{y}) \cdot \mathbf{z}$$

where

$$\mathbf{z} = [I + A]^{n-1}\mathbf{y}$$

But for a given $\mathbf{z}$, $\phi(\mathbf{z})$ is the *largest* number for which the inequality

$$A\mathbf{z} > \phi(\mathbf{z}) \cdot \mathbf{z}$$

holds.

From the last two inequalities and from Section 4.2 we conclude that

$$\phi(\mathbf{z}) \geqslant \phi(\mathbf{y}) \geqslant \phi(\mathbf{x}),$$

so that

$$r = \phi(\mathbf{z}_0) = \max_{\mathbf{x} \geqslant 0} \{\phi(\mathbf{x})\}. \tag{4.22}$$

We now present an argument, very similar to the one in Section 4.2 to show that r, defined by (4.22) is an eigenvalue of A.

We have

$$A\mathbf{z}_0 \geqslant \phi(\mathbf{z}_0) \cdot \mathbf{z}_0 .$$

Assume that

$$A\mathbf{z}_0 \neq \phi(\mathbf{z}_0) \cdot \mathbf{z}_0$$

so that

$$[A - \phi(\mathbf{z}_0)I]\,\mathbf{z}_0 \geqslant \mathbf{0}$$

(at least one element of this vector is $\neq 0$) and

$$[A - \phi(\mathbf{z}_0)I]\,\mathbf{u} > \mathbf{0} \quad \text{(by (4.21))}$$

where

$$\mathbf{u} = [I + A]^{n-1}\mathbf{z}_0 .$$

It follows that

$$A\mathbf{u} > \phi(\mathbf{z}_0)\mathbf{u}.$$

But $\phi(\mathbf{u})$ is the largest number such that

$$A\mathbf{u} \geqslant \phi(\mathbf{u})\mathbf{u},$$

so that

$$\phi(\mathbf{u}) \geqslant \phi(\mathbf{z}_0)$$

which contradicts the definition (4.22).

We conclude that

$$A\mathbf{z}_0 = \phi(\mathbf{z}_0)\mathbf{z}_0$$

or

$$A\mathbf{z}_0 = r\mathbf{z}_0 . \tag{4.32}$$

We are now in a position to discuss the generalised Perron–Frobenius theorem applicable to nonnegative irreducible matrices. Due to the complexity and the length of the proves involved we shall consider the theorem in two parts. The first part is Theorem 4.7, the second part is Theorem 4.8.

Theorem 4.7

Let $A(n \times n) = [a_{ij}]$ be a nonnegative irreducible matrix, then

(1) A has a positive eigenvalue r equal to the spectral radius $\rho(A)$.
(2) The eigenvector associated with the eigenvalue r is positive.
(3) r is a simple eigenvalue of A.

Proof

(1) We have shown above that

$$r = \phi(\mathbf{z}_0) > 0.$$

Let λ be any eigenvalue of A and $\mathbf{u} = [u_1, u_2, \ldots, u_n]'$ the associated normalised eigenvector.

$$\lambda u_i = \sum_j a_{ij} u_j \qquad (i = 1, 2, \ldots, n)$$

so that

$$|\lambda|\,|u_i| = \left|\sum_j a_{ij} u_j\right| \leqslant \sum_j a_{ij}\,|u_j|$$

These inequalities hold for each component of $\mathbf{u}$, including the one which minimises the ratio

$$\frac{\sum_j a_{ij}\,|u_j|}{|u_i|} \qquad \text{where } |u_i| > 0)$$

It follows that

$$|\lambda| \leqslant \min_{x_i>0} \frac{\sum a_{ij} x_j}{x_i} = \phi(\mathbf{x})$$

where

$$\mathbf{x} = [|u_1|,\ |u_2|, \ldots, |u_n|]'.$$

Hence

$$|\lambda| \leqslant \phi(\mathbf{x}) \leqslant \max_{\mathbf{x}>\mathbf{0}} \{\phi(\mathbf{x})\} = r$$

so that

$$r = \rho(A).$$

(Notice that for an irreducible matrix A it is possible that there may be other eigenvalues λ_i such that $|\lambda_i| = r$.)

(2) By (4.23), $\mathbf{z}_0$ is an eigenvector associated with the eigenvalue r. Since

$$\mathbf{z}_0 \in Q,\ \mathbf{z}_0 > \mathbf{0}.$$

(3) By Theorem 4.6,

$$B^m > 0 \quad \text{where} \quad B = [I + A],\ m = n - 1.$$

By Theorem 4.1 (the Perron–Frobenius theorem for positive matrices)

$$\rho(B^m) = (1 + r)^m$$

is a simple eigenvalue of B^m.

By the generalised result (2) above it follows that the multiplicity of the eigenvalue r of A is $k = 1$.

You will have noticed that in proving part (3) of the above theorem we have made use of Theorem 4.1. You may have wondered whether it is possible to make use of this Theorem to prove other properties of non-negative matrices. The following example suggests that this is possible.

Note

Result (3) can also be proved by contradition.

Assume $\rho(A)$ is a multiple eigenvalue of A. Then

$$1 + \rho(A) = \rho(I + A)$$

is a multiple eigenvalue of $[I + A]$. Since

$$\rho[B]^m = \rho[B^m],$$

$\{1 + \rho(A)\}$ is a multiple eigenvalue of $[I + A]^{n-1}$.

But $[I + A]^{n-1} > 0$, so that $1 + \rho(A)$ must be a simple eigenvalue of $[I + A]$ by Theorem 4.1. Hence we must reject the above assumption.

Example 4.7

$A(n \times n)$ is a nonnegative irreducible matrix.

On applying Theorem 4.1, it is found that

$$\rho(B^m) = u^m, u^m > 0$$

and

$$B^m \mathbf{x} = u^m \mathbf{x}, \quad \mathbf{x} > \mathbf{0}$$

where

$$B = [I + A], \quad m = n - 1$$

and

$$u = 1 + r.$$

By considering

$$\mathbf{z} = \sum_{j=0}^{m=1} u^{-j} B^j \mathbf{x}$$

where by definition

$$B^0 = I,$$

show that

(1) r is an eigenvalue of A and $r > 0$.
(2) $\mathbf{z} > \mathbf{0}$ is an eigenvector of A associated with r.

Solution
Since

$$\mathbf{x} > \mathbf{0}, \quad B^j\mathbf{x} > \mathbf{0} \qquad (j = 0, 1, \ldots, m-1)$$

so that $\mathbf{z} > \mathbf{0}$.

$$\begin{aligned} B\mathbf{z} &= \sum_{j=0}^{m-1} u^{-j}B^{j+1}\mathbf{x} \\ &= \sum_{j=0}^{m-2} u^{-j}B^{j+1}\mathbf{x} + u^{1-m}B^m\mathbf{x} \\ &= \sum_{j=1}^{m-1} u^{-j+1}B^j\mathbf{x} + u\mathbf{x} \quad (\text{since } B^m\mathbf{x} = u^m\mathbf{x}) \\ &= u\left[\sum_{j=1}^{m-1} u^{-j}B^j\mathbf{x} + I\mathbf{x}\right] \\ &= u\sum_{j=0}^{m-1} u^{-j}B^j\mathbf{x} \\ &= u\mathbf{z}. \end{aligned}$$

Hence $A\mathbf{z} = r\mathbf{z}$ (see generalisation of results (3)).

Since

$$\mathbf{z} > 0, \quad A\mathbf{z} > \mathbf{0}$$

so that

$$r > 0.$$

Example 4.8
Find the eigenvalues of

$$A = \begin{bmatrix} 0 & 1 & 0 \\ 0 & 0 & 1 \\ 1 & 0 & 0 \end{bmatrix}$$

and the eigenvector $\mathbf{z}_0$ corresponding to $\rho(A)$. Comment on the result.

Solution
The eigenvalues λ_i are

$$1, \frac{1}{2}(-1+\sqrt{3}i), \quad \frac{1}{2}(-1-\sqrt{3}i)$$

$$r = \rho(A) = 1$$

and

$$\mathbf{z}_0 = [1 \ 1 \ 1]'.$$

The point of interest is that $|\lambda_i| = 1$ for all the eigenvalues.

(To conform with the notation used above,

$$\mathbf{z}_0 = \left[\frac{1}{\sqrt{3}} \ \frac{1}{\sqrt{3}} \ \frac{1}{\sqrt{3}}\right]'.\Bigg)$$

Lemma 4.1
Given that A is an irreducible matrix with

$$\rho(A) = r$$

and

$$B(\lambda) = \text{adj}\,[\lambda I - A]$$

then

$$B(r) > 0.$$

Note
In Section 1.8, the adjoint of A was denoted by G.

Proof
Let $C(\lambda)$ be the characteristic polynomial of A.
Then

$$[\lambda I - A]B(\lambda) = C(\lambda)I.$$

Since

$$C(r) = 0$$
$$AB(r) = rB(r)$$

so that each column of $B(r)$ is proportional to the eigenvector $\mathbf{x} > \mathbf{0}$ of A corresponding to r. Hence all entries in each column of $B(r)$ are of the same sign.

The above argument also applies to A' the transpose of A, hence all entries of $B(r)$ are of the same sign.

Also, since $C(\lambda)$ is a product of factors of the form

$$(\lambda - \lambda_i)$$

where

$$r \geqslant |\lambda_i| \quad \text{(all } i\text{)}.$$

$C(r) = 0$, and $C(\lambda) > 0$ for $\lambda > r$.

Hence

$$C'(\lambda) > 0 \quad \text{for} \quad \lambda \geqslant r.$$

In particular

$$C'(r) > 0.$$

But

$$C'(r) = \operatorname{tr} B(r)$$

(see Example 1.7).

The lemma follows. ■

In the beginning of this section reference was made to two irreducible matrices having some similar properties;

$$A_1 = \begin{bmatrix} 0 & 1 & 1 \\ 1 & 1 & 1 \\ 1 & 0 & 1 \end{bmatrix} \quad \text{and} \quad A_2 = \begin{bmatrix} 0 & 2 & 0 \\ 1 & 0 & 1 \\ 0 & 1 & 0 \end{bmatrix}.$$

In fact, there is an important difference between them.

Indeed,

$$A_1^k > 0 \quad \text{for} \quad k \geqslant 2$$

but

$$A_2^k \geqslant 0 \text{ for all } k.$$

Definition

A nonnegative matrix $A(n \times n)$ is called *primitive* if

$$A^k > 0$$

where k is some positive integer.

Since a primitive matrix A is irreducible (see Section 1.12), Theorem 4.7 applies to it. The even stronger result of Theorem 4.2, which was proved for positive matrices, also applies to A.

Indeed if

$$\rho(A) = r,$$

then

$$\rho(A^k) = r^k,$$

and by Theorem 4.2

$$r^k > |\lambda^k| = |\lambda|^k$$

where λ is any other eigenvalue of A.

It follows that

$$r > |\lambda|.$$

This result is certainly borne out in the case of A_1 as indicated at the beginning of this section.

In fact some authors use a classification of the structure of the nonnegative irreducible matrices which depends on the number d, of eigenvalues λ_i such that

$$\rho(A) = |\lambda_i|.$$

If $d = 1$, A is called *primitive*.

If $d > 1$, A is called *imprimitive* or *cyclic.*

d is called the *index of imprimitivity*, or *the index of cyclicity.*

It is to be noted that a positive matrix

$$A > 0$$

is primitive (obviously, since $k = 1$ in this case).

Example 4.9

If $A(n \times n) = [a_{ij}]$ is an irreducible matrix for which

$$a_{ii} > 0 \qquad (i = 1, 2, \ldots, n).$$

Show that A is primitive.

Solution

Let

$$\min_i \{a_{ii}\} = a_{kk} > 0.$$

Then

$$C = \frac{1}{a_{kk}} A = [c_{ij}]$$

A is an irreducible matrix for which

$$c_{ii} \geqslant 1 \quad \text{(all } i\text{)}.$$

Let

$$C = I + B,$$

then

$$C \geqslant \alpha[I+B]$$

for

$$0<\alpha<1.$$

$$C^{n-1} \geqslant \alpha^{n-1}[I+B]^{n-1}>0$$

(by Theorem 4.6).

So

$$A^{n-1}=(a_{kk})^{n-1}C^{n-1}>0.$$

4.5 CYCLIC MATRICES

Having briefly considered primitive matrices we now consider various other aspects of the concept of irreducibility which will lead to a generalisation of the Perron–Frobenius theorem.

We have seen that if A is a primitive matrix

$$A^k>0$$

for some

$$k>0.$$

Also it was pointed out that if A is a cyclic matrix, then

$$A^k \geqslant 0.$$

Note

We are interested in the zero–nonzero pattern of the elements of A^k. We use Boolean operations to simplify the exposition.

We shall now aim to examine and explain the pattern of $A^k (k=1,2,\ldots,)$ which in turn will help us to generalise the theorem.

Consider

$$A=\begin{bmatrix} 0&0&0&1&0&0\\ 1&0&0&0&0&0\\ 0&0&0&1&0&0\\ 0&1&0&0&1&1\\ 1&0&1&0&0&0\\ 0&0&1&0&0&0 \end{bmatrix}$$

Rather than writing out in full the matrices A^2, $A^3,\ldots$ it is sufficient to consider just one of the columns of these matrices.

Notice that if $\mathbf{a}_1^{(r)}$ is the first column of A^r, then $\mathbf{a}_1^{(r+1)}=A\mathbf{a}_1^{(r)}$ is the first column of A^{r+1}.

$$\begin{array}{ccccc} \mathbf{a}_1 & \mathbf{a}_1^{(2)} & \mathbf{a}_1^{(3)} & \mathbf{a}_1^{(4)} & \mathbf{a}_1^{(5)} \\ \begin{bmatrix} 0 \\ 1 \\ 0 \\ 0 \\ 1 \\ 0 \end{bmatrix} & \begin{bmatrix} 0 \\ 0 \\ 0 \\ 1 \\ 0 \\ 0 \end{bmatrix} & \begin{bmatrix} 1 \\ 0 \\ 1 \\ 0 \\ 0 \\ 0 \end{bmatrix} & \begin{bmatrix} 0 \\ 1 \\ 0 \\ 0 \\ 1 \\ 1 \end{bmatrix} & \begin{bmatrix} 0 \\ 0 \\ 0 \\ 1 \\ 0 \\ 0 \end{bmatrix} \end{array}$$

$$\begin{array}{ccccc} \mathbf{a}_4 & \mathbf{a}_4^{(2)} & \mathbf{a}_4^{(3)} & \mathbf{a}_4^{(4)} & \mathbf{a}_4^{(5)} \\ \begin{bmatrix} 1 \\ 0 \\ 1 \\ 0 \\ 0 \\ 0 \end{bmatrix} & \begin{bmatrix} 0 \\ 1 \\ 0 \\ 0 \\ 1 \\ 1 \end{bmatrix} & \begin{bmatrix} 0 \\ 0 \\ 0 \\ 1 \\ 0 \\ 0 \end{bmatrix} & \begin{bmatrix} 1 \\ 0 \\ 1 \\ 0 \\ 0 \\ 0 \end{bmatrix} & \begin{bmatrix} 0 \\ 1 \\ 0 \\ 0 \\ 1 \\ 1 \end{bmatrix} \end{array}$$

It is seen that $\mathbf{a}_j^{(5)}$ is a repeat of $\mathbf{a}_j^{(2)}$ ($j = 1, 4$).

Concentrating on the first columns, this repeat of pattern means that

$$a_{j1}^{(5)} = 1 \quad \text{whenever} \quad a_{j1}^{(2)} = 1.$$

Similarly

$$a_{j1}^{(6)} = 1 \quad \text{whenever} \quad a_{j1}^{(3)} = 1$$

and so on.

More generally, it is seen that

$$a_{j1}^{(3n+2)} = 1 \quad \text{whenever} \quad a_{j1}^{(2)} = 1.$$

Similarly

$$a_{j1}^{(3n)} = 1 \text{ for } n = 2, \ldots \quad \text{whenever} \quad a_{j1}^{(3)} = 1.$$

We can write the above as

$$a_{j1}^{(3n+r)} = 1 \quad \text{whenever} \quad a_{j1}^{(r+3)} = 1$$

for $n = 2, 3, \ldots$ and $r = 0, 1, 2$.

So the pattern is repeated with a *period* $d = 3$. Furthermore, corresponding to $r = 0, 1, \ldots, d-1$, there are subclasses C_r of indices from $\{1, 2, \ldots, 6\}$ defined by

$$C_r = \bigcup_{n>0} \{i : a_{i1}^{(3n+r)} > 0\}.$$

These classes of indices define the nonzero elements in the columns $\mathbf{a}^{(r)}$ which appear periodically (as r is increased) with a period $d = 3$.

In the above example

$$C_1 = \{2, 5, 6\} \quad C_2 = \{4\} \quad \text{and} \quad C_0 = \{1, 3\}.$$

A similar pattern emerges when we consider the fourth column of A^r $(r = 1, 2, \ldots)$.

This time we have

$$C_0' = \{4\} \quad C_1' = \{1, 3\} \quad C_2' = \{2, 5, 6\}.$$

In the general case, where the matrix A is of order $l \times l$ the *index set* of A (see Section 1.3) from which indices $\{i\}$ are selected to form C_r is (of course) $\{1, 2, \ldots, l\}$.

To attempt an explanation of the appearance of the patterns described above, we need to re-examine the concept of matrix irreducibility.

Lemma 4.2

If $A = [a_{ij}]$ is a nonnegative and an irreducible matrix, then for each (i, j), there exists a positive integer k such that

$$a_{ij}^{(k)} > 0.$$

Proof

We make use of a proof by contradiction.

Assume that

$$a_{ij}^{(k)} = 0 \quad (\text{all } k).$$

Let

$$k = u + v.$$

Since

$$A^k = A^u A^v$$

$$a_{ij}^{(k)} = a_{i1}^{(u)} a_{1j}^{(v)} + \ldots + a_{ir}^{(u)} a_{rj}^{(v)} + \ldots a_{in}^{(u)} a_{nj}^{(v)}.$$

Let

$$P = \bigcup_{u>0} \{r : a_{ir}^{(u)} > 0\}.$$

and

$$Q = \bigcup_{v>0} \{r : a_{ir}^{(v)} > 0\}.$$

Since

$$a_{ij}^{(k)} = 0, \quad P \cap Q = \phi.$$

As the matrix A cannot have either a row or a column of zero elements

$$P \neq \phi \quad \text{and} \quad Q \neq \phi.$$

This result leads to a contradiction.

Since $P \neq \phi$, $\exists$ an integer $p \in P$ such that

$$a_{pj}^{(l)} > 0.$$

Since

$$Q \neq \phi, \qquad \exists \text{ an integer } q \in Q$$

such that

$$a_{iq}^{(m)} > 0$$

for some $l, m \in Z^+$.

It follows that

$$0 < a_{iq}^{(m)} a_{pj}^{(l)} \leqslant a_{ij}^{(l+m)}.$$

This contradicts the assumption and the result follows.

Corollary 1

Since $A^{m+1} = AA^m$

$$a_{rj}^{(m+1)} \geqslant a_{rs} a_{sj}^{(m)}.$$

If

$$r \notin P, \quad a_{rj}^{(m+1)} = 0.$$

If

$$s \in P, \quad a_{sj}^{(m)} > 0.$$

It follows that

$$a_{rs} = 0$$

if

$$s \in P, \quad r \notin P.$$

Corollary 2

Since $a_{ij}^{(k)} > 0$ for some k, it follows that

$$A + A^2 + \ldots + A^r > 0$$

for some $r \geqslant 1$. ∎

Having made a number of observations relating to the pattern of zero and non-zero elements of powers of a cyclic matrix, we shall now prove some of these results.

Lemma 4.3

If $A = [a_{ij}]$ is an irreducible matrix, then

$$a_{ii}^{(kd)} > 0$$

for all $k \geqslant n_0 (k, n_0 \in Z^+)$ where

$$d = \text{g.c.d. of } S = \{r : a_{ii}^{(r)} > 0\}.$$

Proof

By Lemma 2 $\exists\ m, n \in S$.

But

$$a_{ii}^{(m+n)} \geqslant a_{ii}^{(m)} a_{ii}^{(n)} > 0.$$

So that

$$(m+n) \in S.$$

Therefore S is closed under addition.

It is shown in Appendix I that if $d =$ g.c.d. S then $kd \in S$ for all sufficiently large k.

Hence

$$a_{ii}^{(kd)} > 0.$$

Lemma 4.4

For a fixed j (that is a fixed column of A) and each i from $\{1, 2, \ldots, n\}$ there is an integer r_i $(0 \leqslant r_i < d)$ such that

(1) if $a_{ij}^{(m)} > 0$ then $m = qd + r_i = r_i \pmod d$ (where q is an integer), and

(2) $a_{ij}^{(kd+r_i)} > 0$ for $k \geqslant K$ (K is some positive integer).

Proof

(1) By Lemma 4.2, $\exists s, t \in Z^+$ such that

$$a_{ij}^{(s)} > 0 \quad \text{and} \quad a_{ji}^{(t)} > 0.$$

Now

$$a_{ii}^{(m+t)} \geqslant a_{ij}^{(m)} a_{ji}^{(t)} > 0$$

and

$$a_{ii}^{(s+t)} \geqslant a_{ij}^{(s)} a_{ji}^{(t)} > 0.$$

By Lemma 4.3, d divides both $(m+t)$ and $(s+t)$, hence d divides the difference

$$(m+t) - (s+t) = m - s.$$

It follows that

$$m - s = 0 \pmod{d}$$

or

$$m = r_i \pmod{d}.$$

(2) By (1) above, $\exists q \in Z^+$ such that

$$a_{ij}^{(qd+r_i)} > 0.$$

By Lemma 4.3

$$a_{ii}^{(pd)} > 0 \quad \text{for} \quad p \geqslant n_0.$$

Let $k = q + p$ so that $k \geqslant K = n_0 + q$ then

$$kd + r_i = qd + pd + r_i$$

It follows that

$$a_{ij}^{(kd+r_i)} \geqslant a_{ii}^{(qd)} \, a_{ij}^{(pd+r_i)} > 0$$

for all $k \geqslant K$. ∎

We can now define formally the subclasses C_r of indices in the same residue class (mod d). Although the definition focuses on the first column of the matrices A^k $(k = 1, 2, \ldots)$, this is an arbitrary choice. It will be shown that a similar definition applies to the indices associated with any fixed column of these matrices.

Definition
Given a nonnegative irreducible matrix $A(l \times l)$ the set of indices

$$C_r = \bigcup_{n>0} \left\{ i : a_{i1}^{(nd+r)} > 0 \right\} \qquad 0 \leqslant r < d$$

is called a *subclass* of the class $\{1, 2, \ldots, l\}$.

Lemma 4.5
The following properties apply to the subclasses C_r defined above:

(1) $\bigcup_{r=0}^{d-1} C_r = \{1, 2, \ldots, l\}$

(2) $C_r \neq \phi \; (r = 1, 2, \ldots, d-1)$

(3) $C_r \cap C_s = \phi \; (r, s = 1, 2, \ldots, d-1; r \neq s)$.

Proof
(1) follows immediately from Lemma 4.4 (2).
(2) Assume that $C_k = \phi$.

Then

$$a_{i1}^{(dn+k)} = 0 \quad \text{for} \quad i = 1, 2, \ldots, l.$$

So the first column of A^m, $m = dn + k$, is $\mathbf{0}$, hence A is reducible. But since A is irreducible, the assumption is false.

(3) Assume $k \in C_r \cap C_s$, so that

$$a_{k1}^{(pd+r)} > 0 \quad \text{and} \quad a_{k1}^{(qd+s)} > 0.$$

By (1) of Lemma 4.4

$$pd + r = r_k \pmod d$$

$$qd + s = r_k \pmod d.$$

Hence

$$r = s = r_k.$$

We can now verify the fact, already noted in the beginning of this section, that for each column of A the composition of the subclasses $\{C_i\}$ remains the same although their order may be subject to a cyclic permutation.

Lemma 4.6

Let

$$C_r' = \bigcup_{m>0} \{i; a_{iq}^{(md+r')} > 0\}$$

where q is a fixed index from the set $\{1, 2, \ldots, l\}$. Then $\{C_r'\}$ is a cyclic permutation of $\{C_r\}$.

Proof

Since a_{q1} is in the first column of A

$$a_{q1}^{(md+r'')} > 0 \qquad \text{for some } 0 \leqslant r'' < d \text{ (by Lemma 4.5 (2)).}$$

Also

$$a_{iq}^{(kd+r')} > 0 \quad \text{for some } 0 < r' < d \text{ and } k \leqslant K \text{ (by Lemma 4.4 (2)).}$$

Hence

$$a_{i1}^{(kd+r'+md+r'')} \geqslant a_{iq}^{(kd+r')} a_{q1}^{(md+r'')} > 0.$$

But a_{i1} (for $i = 1, 2, \ldots, l$) are the elements of the first column of A and so are classified in $\{C_r\}$, hence

$$kd + r' + md + r'' = r \pmod d$$

or

$$r' + r'' = r \pmod d$$

where r' defines the subclasses $\{C_r'\}$ with reference to the qth column of A.

To illustrate this result, consider $q = 4$ for the matrix A discussed at the beginning of this section.

We have

$$a_{14}^{(2)} \; (= a_{14}^{(5)}) > 0,$$

hence

$$r'' = 2.$$

So

$$r' + 2 = r \,(\text{mod } 3)$$

where r defines the subclasses $\{C_r\}$ relative to $\mathbf{a}_1$.

They are

$$C_0 = \{1, \; 3\}, \quad C_1 = \{2, \; 5, \; 6\}, \quad \text{and} \quad C_2 = \{4\}.$$

For

$$r = 0, 1, 2, \quad r' = 1, 2, 0$$

hence

$$C_0' = C_2 = \{4\}, \quad C_1' = C_0 = \{1, \; 3\} \quad \text{and} \quad C_2' = C_1 = \{2, \; 5, \; 6\}$$

which is precisely the result we had found already. ■

It is worthwhile to take a short pause at this stage to look back at what has been demonstrated in this section.

We have shown that associated with a nonnegative irreducible matrix A, is a number d called the period of A.

By Lemma 4.3, if $d = 1$

$$a_{ii}^{(k)} > 0 \quad \text{for all} \quad k \geqslant n_0.$$

Hence, by a generalisation of Example 4.9, $A^m > 0$, so that A is primitive. On the other hand, if $d > 1$

$$A^m \geqslant 0 \quad \text{(see Lemma 4.3)}$$

and so the matrix A is cyclic.

For a fixed column of the successive powers of a cyclic matrix A, the positive entries occur in d subclasses C_r.

Lemma 4.7

If $a_{rs} > 0$ and $s \in C_j$, then $r \in C_{j+1}$ for $j = 0, 1, \ldots, d-1$.

Proof

Since $s \in C_j$, whenever $a_{s1}^{(m)} > 0$

$$s = nd + j \qquad \text{(by definition of } C_r\text{)}.$$

Assume

$$a_{s1}^{(m)} > 0.$$

Since

$$a_{r1}^{(m+1)} \geqslant a_{rs}\, a_{s1}^{(m)}$$

then

$$(m + 1) = nd + (j + 1)$$

so

$$r \in C_{j+1}.$$

Corollary
If $A = [a_{ij}]$ is a cyclic matrix, then

$$a_{ii} = 0 \qquad (i = 1, 2, \ldots, l).$$

Proof
Assume that $a_{kk} > 0$.
By Lemma 4.7,

$$k \in C_r \quad \text{and} \quad k \in C_{r+1},$$

but

$$C_r \cap C_{r+1} = \phi.$$

Hence

$$a_{kk} = 0. \quad ■$$

Having achieved a partitioning of the index set $\{1, 2, \ldots, l\}$ associated with A into subclasses C_r $(r = 0, 1, \ldots, d - 1)$, we are now ready to proceed to the next stage along our goal of generalising the Perron–Frobenius theorem to cyclic matrices.

The relabelling of the indices of any square matrix A is equivalent to a simultaneous permutation of the rows and columns of the matrix. We have seen in Section 1.12 that this is achieved by a transformation of the form

$$PAP^{-1}.$$

This is a similarity transformation of A (see [B7]);

(1) it leaves the eigenvalues unchanged, and
(2) the powers of A are similarly transformed, so that

$$(PAP^{-1})^k = PA^kP^{-1}.$$

We will relabel the indices of a cyclic matrix by taking, in their natural order, the indices in the subclass C_r followed by the indices in C_{r+1} $(r = 1, 2, \ldots, d-1)$ until the subclasses are exhausted. In this way the resulting form of A is particularly convenient as it will allow us to examine its eigenvalues in terms of primitive matrices.

As an illustrative example of a transformation described above, we consider once again the matrix discussed at the beginning of this section.

It is usual to relabel the indices in two equivalent forms as;

$$\{C_1 C_2 C_0\} \quad \text{or} \quad \{C_0 C_1 C_2\}.$$

We shall choose the first form, so that the matrix A is transformed into:

$$\begin{array}{cc} & \begin{array}{ccc:c:cc} \multicolumn{3}{c}{\overbrace{\quad}^{C_1}} & \overbrace{}^{C_2} & \multicolumn{2}{c}{\overbrace{\quad}^{C_0}} \\ 2 & 5 & 6 & 4 & 1 & 3 \end{array} \\ \begin{array}{r} C_1 \left\{ \begin{array}{c} 2 \\ 5 \\ 6 \end{array} \right. \\ C_2 \{ \; 4 \\ C_0 \left\{ \begin{array}{c} 1 \\ 3 \end{array} \right. \end{array} & \left[\begin{array}{ccc:c:cc} 0 & 0 & 0 & 0 & 1 & 0 \\ 0 & 0 & 0 & 0 & 1 & 1 \\ 0 & 0 & 0 & 0 & 0 & 1 \\ \hdashline 1 & 0 & 0 & 0 & 0 & 0 \\ \hdashline 0 & 0 & 0 & 1 & 0 & 0 \\ 0 & 0 & 0 & 1 & 0 & 0 \end{array} \right] \end{array}$$

Notation

We write $n(C_r)$ to denote the number of indices in C_r. In what follows we shall denote the subclass C_0 by C_3 whenever such notation is particularly meaningful.

The transformed matrix can be partitioned into blocks by lines separating the indices in the subclasses $\{C_r\}$.

Lemma 4.7 now proves its worth since it can be used to determine the possible type of element in each block.

(1) The diagonal blocks are zero matrices of order $n(C_r) \times n(C_r)$.

(2) The blocks which contain non-zero elements are defined by the intersection of the indices in:

$$C_2 \times C_1, \; C_3 \times C_2 \quad \text{and} \quad C_1 \times C_0.$$

(3) The remaining blocks, defined by the intersection of the indices in

$$C_1 \times C_2, \; C_2 \times C_3 \quad \text{and} \quad C_0 \times C_1$$

are all zero matrices.

So the matrix A has been transformed into the form

$$PAP^{-1} = \begin{bmatrix} 0 & 0 & A_3 \\ A_1 & 0 & 0 \\ 0 & A_2 & 0 \end{bmatrix}$$

where

$$A_1 = [1 \ \ 0 \ \ 0], \quad A_2 = \begin{bmatrix} 1 \\ 1 \end{bmatrix} \quad \text{and} \quad A_3 = \begin{bmatrix} 1 & 0 \\ 1 & 1 \\ 0 & 1 \end{bmatrix}$$

where A_r is of order $n(C_{r+1}) \times n(C_r)$.

(A_3 can and often is denoted by A_0.)

The construction of the permutation P is very simple following the discussion in Section 1.12. The rows of P are the unit vectors $\{e_i\}$ where $\{i\}$ is the sequence of relabelled indices starting with C_1 and ending with C_0. For the above example

$$P = \begin{bmatrix} 0 & 1 & 0 & 0 & 0 & 0 \\ 0 & 0 & 0 & 0 & 1 & 0 \\ 0 & 0 & 0 & 0 & 0 & 1 \\ 0 & 0 & 0 & 1 & 0 & 0 \\ 1 & 0 & 0 & 0 & 0 & 0 \\ 0 & 0 & 1 & 0 & 0 & 0 \end{bmatrix}$$

The convenience of this form of PAP^{-1} becomes evident when it is raised to the power $d (d = 2$ for this example).

$$PA^3P^{-1} = \begin{bmatrix} Q_3 & 0 & 0 \\ 0 & Q_2 & 0 \\ 0 & 0 & Q_1 \end{bmatrix}$$

where

$$Q_3 = A_3 A_2 A_1$$

$$Q_2 = A_1 A_3 A_2$$

$$Q_1 = A_2 A_1 A_3$$

and Q_r is of order $n(C_r) \times n(C_r)$.

The significance of this result is not only due to PA^3P^{-1} being in a block diagonal form, but also due to the fact that the diagonal matrices Q_1, Q_2 and Q_3 are square and primitive.

To prove that the matrices Q_r are primitive is not difficult. Since the period of A is $d = 3$

$$a_{ii}^{(3k)} > 0 \quad (\text{all } i) \text{ for } k \geqslant K.$$

So the diagonal elements of

$$Q_r^k \ (r = 1, 2, 3)$$

are all positive for $k \geqslant K$.

We can now show that Q_1 is primitive.

If

$$s \in C_0,$$

then

$$a_{1s}^{(3q)} > 0 \quad \text{for some } q \in Z^+ \quad (\text{by Lemma 4.3 (1)})$$

but

$$a_{ss}^{(3k)} > 0 \quad \text{for all } k \geqslant K,$$

so that

$$a_{1s}^{(3(k+q)} \geqslant a_{1s}^{(3q)} a_{ss}^{(3k)} > 0$$

for all sufficiently large $(k + q)$.

So

$$a_{1s}^{(3m)} > 0 \quad \text{for } all\ m = k + q \geqslant N.$$

If

$$r \in C_0,$$

then

$$A_{r1}^{(3l)} > 0, \quad \text{for some } l \in Z^+.$$

By an argument similar to the above

$$a_{r1}^{(3n)} > 0 \quad \text{for all } n \geqslant N_1.$$

It follows that

$$a_{rs}^{(3(m+n))} \geqslant a_{r1}^{(3n)} a_{1s}^{(3m)} > 0$$

for all sufficiently large $(m + n)$, so Q_1 is primitive.

Q_2 and Q_3 can also be shown to be primitive by either a similar method or deduced from the above result.

We can now consider the eigenvalues of the Q_i matrices and show that each Q_i has the same spectral radius $\tilde{r} = r^d$, where $r = \rho(A)$.

Since Q_3 and Q_2 are primitive $\exists\, \mathbf{x} > \mathbf{0}$, and $\mathbf{y} > \mathbf{0}$ such that

$$Q_3\mathbf{x} = r_3\mathbf{x} \quad \text{and} \quad Q_2\mathbf{y} = r_2\mathbf{y}$$

where

$$r_3 = \rho(Q_3) \quad \text{and} \quad r_2 = \rho(Q_2).$$

But

$$Q_3 = A_1A_3A_2A_1 = Q_2A_1$$

and

$$A_1Q_3\mathbf{x} = r_3(A_1\mathbf{x})$$

so that

$$Q_2(A_1\mathbf{x}) = r_3(A_1\mathbf{x})$$

which shows that r_3 is an eigenvalue of Q_2 and

$$A_1\mathbf{x} > \mathbf{0}$$

is an associated eigenvector.

From the discussion in the previous section

$$r_3 = r_2.$$

By a similar argument we conclude that

$$r_3 = r_1$$

where

$$r_1 = \rho(Q_1).$$

As

$$\rho(PA^3P^{-1}) = \rho(A^3) = r^3$$

where

$$r = \rho(A)$$

it follows that

$$\tilde{r} = r_1 = r_2 = r_3 = r^3.$$

So r is the positive cube root of $\tilde{r}$ (by Theorem 4.7).

Every eigenvalue λ of A such that

$$\lambda^3 = r^3$$

is called a *dominant* root and must be of the form

$$\lambda = r \exp i \left\{\frac{2\pi k}{3}\right\} \qquad k = 0, 1, 2.$$

Example 4.10

Illustrate the results proved in this section when

$$A = \begin{array}{c} \\ 1 \\ 2 \\ 3 \\ 4 \\ 5 \\ 6 \end{array} \begin{array}{c} \begin{array}{cccccc} 1 & 2 & 3 & 4 & 5 & 6 \end{array} \\ \begin{bmatrix} 0 & 2 & 0 & 0 & 0 & 0 \\ 0 & 0 & 0 & 1 & 0 & 0 \\ 0 & 0 & 0 & 2 & 0 & 1 \\ 2 & 0 & 0 & 0 & 1 & 0 \\ 0 & 1 & 1 & 0 & 0 & 0 \\ 1 & 0 & 0 & 0 & 1 & 0 \end{bmatrix} \end{array}$$

Solution

Note
To determine the subclasses of the index set we can use Boolean operations. For calculations of eigenvalues we must use the normal matrix operations.

$$\mathbf{a}_1 \quad \mathbf{a}_1^{(2)} \quad \mathbf{a}_1^{(3)} \quad \mathbf{a}_1^{(4)}$$
$$\begin{bmatrix} 0 \\ 0 \\ 0 \\ 1 \\ 0 \\ 1 \end{bmatrix} \begin{bmatrix} 0 \\ 1 \\ 1 \\ 0 \\ 0 \\ 0 \end{bmatrix} \begin{bmatrix} 1 \\ 0 \\ 0 \\ 0 \\ 1 \\ 0 \end{bmatrix} \begin{bmatrix} 0 \\ 0 \\ 0 \\ 1 \\ 0 \\ 1 \end{bmatrix}$$

Hence
$$C_1 = \{4, 6\}, \quad C_2 = \{2, 3\}, \quad C_0 = \{1, 5\}.$$

So

$$PAP^{-1} = \begin{array}{c} \left\{\begin{array}{c} 4 \\ 6 \end{array}\right. \\ \left\{\begin{array}{c} 2 \\ 3 \end{array}\right. \\ \left\{\begin{array}{c} 1 \\ 5 \end{array}\right. \end{array} \begin{array}{c} \overbrace{\begin{array}{cc} 4 & 6 \end{array}}^{} \quad \overbrace{\begin{array}{cc} 2 & 3 \end{array}}^{} \quad \overbrace{\begin{array}{cc} 1 & 5 \end{array}}^{} \\ \begin{bmatrix} 0 & 0 & 0 & 0 & 2 & 1 \\ 0 & 0 & 0 & 0 & 1 & 1 \\ \hdashline 1 & 0 & 0 & 0 & 0 & 0 \\ 2 & 1 & 0 & 0 & 0 & 0 \\ \hdashline 0 & 0 & 2 & 0 & 0 & 0 \\ 0 & 0 & 1 & 1 & 0 & 0 \end{bmatrix} \end{array}$$

Hence

$$A_1 = \begin{bmatrix} 1 & 0 \\ 2 & 1 \end{bmatrix}, \quad A_2 = \begin{bmatrix} 2 & 0 \\ 1 & 1 \end{bmatrix} \quad \text{and} \quad A_3 = \begin{bmatrix} 2 & 1 \\ 1 & 1 \end{bmatrix}.$$

So that

$$Q_3 = \begin{bmatrix} 7 & 1 \\ 5 & 1 \end{bmatrix}, \quad Q_2 = \begin{bmatrix} 5 & 1 \\ 13 & 3 \end{bmatrix} \quad \text{and} \quad Q_1 = \begin{bmatrix} 4 & 2 \\ 7 & 4 \end{bmatrix}.$$

As

$$|\lambda I - Q_i| = \lambda^2 - 8\lambda + 2 \qquad (i = 1, 2, 3)$$

the Q_i have equal eigenvalues.

$$\tilde{r} = \rho(Q_i) = 4 + \sqrt{14} = 7.7417.$$

It follows that

$$r = \rho(A) = 1.978.$$

The other two dominant eigenvalues are

$$\lambda_2 = 1.978 \exp i \left\{\frac{2\pi}{3}\right\} \quad \text{and} \quad \lambda_3 = 1.978 \exp i \left\{\frac{4\pi}{3}\right\}$$

or

$$\lambda_2 = 1.978\left[-\frac{1}{2} + i\frac{\sqrt{3}}{2}\right], \quad \lambda_3 = 1.978\left[-\frac{1}{2} - i\frac{\sqrt{3}}{2}\right]$$

We have proved in this section a number of results which are characteristic of irreducible cyclic matrices. It may be of some interest to stress two fairly obvious observations.

(1) The properties of irreducible matrices as summarised in Theorem 4.7 apply to cyclic matrices.
(2) The proofs of results in this section were not in every case the most general ones. It was considered that the understanding of the fairly complicated structure of cyclic matrices was more important than the presentation of rigorous proofs which can be found in the literature mentioned in the Bibliography in this book.

We now summarise the results which form the second part of the Perron–Frobenius theorem for irreducible matrices. It will be appreciated in what follows that the matrix A is nonnegative and irreducible and that if

$$d = 1,$$

A is primitive and if

$$d > 1,$$

A is cyclic.

Theorem 4.8
If A is an irreducible matrix of order $(n \times n)$

(1) There exists a permutation matrix P and an integer $d \geqslant 1$ such that

$$PAP^{-1} = \begin{bmatrix} 0 & 0 & 0 \dots 0 & 0 & A_d \\ A_1 & 0 & 0 \dots 0 & 0 & 0 \\ 0 & A_2 & 0 \dots 0 & 0 & 0 \\ \vdots & & & & \\ 0 & 0 & 0 \dots A_{d-2} & 0 & 0 \\ 0 & 0 & 0 \dots 0 & A_{d-1} & 0 \end{bmatrix}$$

where the diagonal blocks are square zero matrices.

(2) PA^dP^{-1} is a direct sum

$$Q_1 \oplus Q_2 \oplus \dots \oplus Q_d$$

of nonnegative matrices Q_i.

(3) Q_i are primitive matrices each with the same maximal eigenvalue.

$$\rho(Q_i) = r^d \quad \text{where } r = \rho(A). \qquad (i = 1, 2, \dots, d)$$

(4) The dominant roots λ of A, are the distinct roots of

$$\lambda^d - r^d = 0.$$

Proof
The proofs of the four parts of the theorem have been discussed above. In fact, for a cyclic matrix A we can say more about its eigenvalues. Using the results in Section 4.4 (also Section 1.7) we can determine the other eigenvalues of A. We know that all the dth roots of the real eigenvalues of the Q_i are eigenvalues of A. The following lemma is useful in that it allows us to evaluate $(d - 1)$ eigenvalues for each known eigenvalue of A. For example, having found the maximal eigenvalue r, the lemma can be used to determine the remaining dominant eigenvalues of A.

Lemma 4.8
If λ is any eigenvalue of A then

$$\lambda\sigma$$

where

$$\sigma^d = 1$$

is also an eigenvalue of A.

Proof
To simplify the proof we begin with a matrix A of order $(N \times N)$ assumed to have a period $d = 3$, so that

$$T = PAP^{-1} = \begin{array}{c} \\ C_1 \\ C_2 \\ C_0 \end{array} \begin{array}{c} \begin{array}{ccc} C_1 & C_2 & C_0 \end{array} \\ \begin{bmatrix} 0 & 0 & A_0 \\ A_1 & 0 & 0 \\ 0 & A_2 & 0 \end{bmatrix} \end{array} \tag{4.24}$$

where the C_i indicate the orders of the blocks in the partitioned matrix T. Let λ be any eigenvalue of T and $\mathbf{x}$ the corresponding eigenvector so that

$$T\mathbf{x} = \lambda\mathbf{x} \tag{4.25}$$

We can partition $\mathbf{x}$ into the form

$$\mathbf{x}' = [\mathbf{x}_1' \ \ \mathbf{x}_2' \ \ \mathbf{x}_0']$$

so that (4.25) becomes

$$T\begin{bmatrix} \mathbf{x}_1 \\ \mathbf{x}_2 \\ \mathbf{x}_0 \end{bmatrix} = \lambda \begin{bmatrix} \mathbf{x}_1 \\ \mathbf{x}_2 \\ \mathbf{x}_0 \end{bmatrix}$$

The orders of the vectors $\mathbf{x}_1, \mathbf{x}_2$ and $\mathbf{x}_0$ are respectively $\mathrm{n}(C_1)$, $\mathrm{n}(C_2)$ and $\mathrm{n}(C_0)$.
We can assume that the components of these vectors are as follows:

$$\mathbf{x}_1 = [x_{11}x_{12} \dots x_{1\mathrm{n}(C_1)}]'$$

$$\mathbf{x}_2 = [x_{21}x_{22} \dots x_{2\mathrm{n}(C_2)}]'$$

$$\mathbf{x}_0 = [x_{01}x_{02} \dots x_{0\mathrm{n}(C_0)}]'$$

We define the following new vectors, each of order N:

$$\mathbf{z}_1 = [x_{11}, x_{12}, \dots, x_{1\mathrm{n}(C_1)} \underbrace{0 \ 0 \ \dots \ 0}_{N-\mathrm{n}(C_1)}]' \tag{4.26}$$

$$\mathbf{z}_2 = [\underbrace{0 \ 0 \ \dots \ 0}_{\mathrm{n}(C_1)} \ x_{21}, x_{22}, \dots, x_{2\mathrm{n}(C_2)} \underbrace{0 \ 0 \ \dots \ 0}_{\mathrm{n}(C_0)}]'$$

$$\mathbf{z}_0 = [\underbrace{0 \ 0 \ \dots \ 0}_{N-\mathrm{n}(C_0)} \ x_{01}, x_{02}, \dots, x_{0\mathrm{n}(C_0)}]'$$

Note that from the definitions of $\mathbf{z}_i$

$$\sum \mathbf{z}_i = \mathbf{x} \quad \text{(an eigenvector of } T\text{)}$$

Taking into account (4.24) and (4.26), we obtain from (4.25)

$$T\mathbf{z}_0 = \lambda\mathbf{z}_1,\ T\mathbf{z}_1 = \lambda\mathbf{z}_2 \quad \text{and} \quad T\mathbf{z}_2 = \lambda\mathbf{z}_0.$$

More generally, when A has period d, we obtain

$$T\mathbf{z}_i = \lambda\mathbf{z}_{i+1} \qquad (i = 0, 1, \ldots, d-1) \tag{4.27}$$

where

$$\mathbf{z}_{\mathrm{d}} = \mathbf{z}_0.$$

We next use the technique introduced in Example 4.7.

Let σ be any dth root of unity, so that

$$\sigma^{\mathrm{d}} = 1.$$

Consider

$$\mathbf{y} = \sum_{k=0}^{d-1} \sigma^{-k}\mathbf{z}_k.$$

Then

$$T\mathbf{y} = \lambda \sum_{k=0}^{d-1} \sigma^{-k}\mathbf{z}_{k+1} \qquad \text{(by 4.27)}$$

$$= \lambda\left[\sum_{k=0}^{d-2} \sigma^{-k}\mathbf{z}_{k+1} + \sigma^{d-1}\mathbf{z}_{\mathrm{d}}\right]$$

$$= \lambda\left[\sum_{k=1}^{d-1} \sigma^{-k+1}\mathbf{z}_k + \sigma\mathbf{z}_0\right]$$

$$= \lambda\sigma\left[\sum_{k=0}^{d-1} \sigma^{-k}\mathbf{z}_k\right]$$

$$= \lambda\sigma\mathbf{y}$$

It follows that $(\lambda\sigma)$ is an eigenvalue of T and hence of A.

Example 4.11

Find all the eigenvalues of the matrix A considered in Example 4.10.

Solution

For this matrix A it was found that

$$d = 3$$

and that the eigenvalues of Q_i $(i = 1, 2, 3)$ are

$$\mu_1 = 4 + \sqrt{14} \quad \text{and} \quad \mu_2 = 4 - \sqrt{14}.$$

Hence two of the eigenvalues of A are

$$\lambda_1 = r = [4 + \sqrt{14}]^{1/3}$$

and

$$\lambda_2 = [4 - \sqrt{14}]^{1/3}$$

Let σ_i $(i = 0, 1, 2)$ be the cube roots of 1, that is

$$\sigma_0 = 1, \quad \sigma_1 = \exp i \left\{\frac{2\pi}{3}\right\} \quad \text{and} \quad \sigma_2 = \exp i \left\{\frac{4\pi}{3}\right\} = \sigma_1^2$$

By Lemma 4.8, the six eigenvalues of A are

$$\lambda_1, \lambda_1\sigma_1, \lambda_1\sigma_2, \lambda_2, \lambda_2\sigma_1 \text{ and } \lambda_2\sigma_2. \quad ■$$

From the above discussion we note the following:

(1) The spectrum of the matrices

$$A \text{ and } \sigma A \text{ where } \sigma = \exp i \left\{\frac{2\pi}{d}\right\}$$

are the same.

(2) Let

$$\hat{\sigma} = \exp i \left\{\frac{2\pi}{t}\right\},$$

then A and $\hat{\sigma}A$ have the same spectrum if $t = d$.

(3) If $t > d$ then

$$\sigma > \hat{\sigma}.$$

So the set $\{\lambda_j \hat{\sigma}^i\}$ has more elements than $\{\lambda_j \sigma^i\}$ (where λ_j is an eigenvalue of A). This contradicts Lemma 4.7.

It follows that A and $\hat{\sigma}A$ do not have the same spectrum.

Lemma 4.9

Let $A(n \times n)$ be a cyclic matrix of period d having eigenvalues $\lambda_1, \lambda_2, \ldots, \lambda_n$. The points in the complex plane corresponding to $\lambda_1, \lambda_2, \ldots, \lambda_n$ are invariant under a rotation through an angle $2\pi/d$.

Proof

This is a direct result of Lemma 4.8 and is well illustrated by the results obtained in Example 4.11. ■

Rotating the six eigenvalues through an angle

$$\frac{2\pi}{3}$$

is equivalent to multiplying each λ_i by

$$\exp i\left\{\frac{2\pi}{3}\right\} = \sigma_1 .$$

The six specified numbers become

$$\lambda_1\sigma_1, \lambda_1\sigma_2, \lambda_1, \lambda_2\sigma_1, \lambda_2\sigma_2 \text{ and } \lambda_2 .$$

This is the same set of eigenvalues as evaluated in Example 4.11.

Lemma 4.10

Let $A(n \times n)$ be an irreducible cyclic matrix with a period d and a characteristic equation

$$C(\lambda) = \lambda^n + a_1\lambda^{n_1} + a_2\lambda^{n_2} + \ldots + a_k\lambda^{n_k}$$

where

$$n > n_1 > n_2 > \ldots > n_k$$

and

$$a_i \neq 0 \qquad (i = 1, 2, \ldots, k).$$

Let

$$S = \{n - n_1 \; n - n_2, \ldots, n - n_k\}$$

then

$$d = \text{g.c.d.}(S).$$

Proof

Let $\sigma = \exp i\{2\pi/d\}$ and $\hat{\sigma} = \exp i\{2\pi/t\}$ $(t \in \mathbf{Z}^+)$.

We consider the integers $\{t\}$ such that

$$A \text{ and } \hat{\sigma}A$$

have the same eigenvalues.

Since the characteristict equations of A and $\hat{\sigma}A$ are equal

$$\lambda^n + a_1\lambda^{n_1} + \ldots + a_k\lambda^{n_k} = (\hat{\sigma}\lambda)^n + a_1(\hat{\sigma}\lambda)^{n_1} + \ldots + a_k(\hat{\sigma}\lambda)^{n_k}$$

hence

$$a_j = a_j\,\hat{\sigma}^{(n-n_j)} \qquad (j = 1, 2, \ldots, k).$$

This is only possible when

$$\hat{\sigma}^{(n-n_j)} = 1 \qquad (j = 1, 2, \ldots, k) \tag{4.28}$$

that is if t is a common divisor of S.

We claim that

$$d = t_1 = \text{g.c.d.}(S).$$

Assume on the contrary that

$$d = t$$

Then both t and t_1 satisfy (4.28), and

$$t_1 \geqslant t = d.$$

If $t_1 > t = d$, then

$$\hat{\sigma} > \sigma$$

and as we have seen, it is not possible in this case for A and $\hat{\sigma}A$ to have the same eigenvalues. It follows that

$$d = t_1 = \text{g.c.d.}\{S\}. \quad \blacksquare$$

In Example 4.9 we have noted a sufficient condition for an irreducible matrix A to the primitive. We can deduce a necessary condition for A to be cyclic from Lemma 4.10.

Example 4.12
Show that a necessary condition for an irreducible matrix $A(n \times n)$ to be cyclic is that $\text{tr}(A) = 0$.

Solution

$$\text{tr}(A) = \sum_{i=1}^{n} \lambda_i = \text{coefficient of } \lambda^{n-1}$$

in the characteristic equation $C(\lambda)$ of A.

The condition now follows firm Lemma 4.10.

Conversely, if $\text{tr}(A) > 0$, then

$$n - n_1 = 1$$

hence

$$d = 1$$

and A is primitive.

(This is a stronger condition than the one in Example 4.9.)

Note
Remember, since we are dealing with a nonnegative matrix A

$$\text{tr}(A) \geqslant 0.$$

4.6 REDUCIBLE MATRICES

By definition (1.43) the matrix $A(n \times n) \geqslant 0$ is reducible if its rows and columns can be rearranged simultaneously so that the resulting matrix has the form

$$\begin{bmatrix} A_{11} & A_{12} \\ 0 & A_{22} \end{bmatrix}$$

where A_{11} and A_{22} are square matrices.

If A_{11} or A_{22} are reducible we continue the above process until finally we can construct a permutation matrix P such that

$$PAP' = \begin{bmatrix} B_{11} & & & \text{X} \\ & B_{22} & & \\ & & \ddots & \\ \text{O} & & & B_{mm} \end{bmatrix} \tag{4.29}$$

where O represent zero entries, X represents nonnegative entries, and

$$B_{jj} \; (j = 1, 2, \ldots, m)$$

is an irreducible matrix.

Notice that B_{jj} may be of order (1×1), the entry being nonnegative (possibly zero).

(4.28) is called a *normal form* of A.

Some important properties of A can be derived from those of the B_{jj}. For example, the eigenvalues of A are the eigenvalues of the B_{jj} $(j = 1, 2, \ldots, m)$.

Theorem 4.9

If $A(n \times n) = [a_{ij}]$ is a nonnegative matrix then

(1) A has an eigenvalue $r \geqslant 0$, where $r = \rho(A)$.
(2) Corresponding to the eigenvalue r, the eigenvector $\mathbf{x} \geqslant \mathbf{0}$.
(3) r does not decrease when an entry in A is increased.

(Notice

(1) $\mathbf{x}$ may have some zero components.
(2) r does not necessarily increase when an entry in A is increased.
(3) It is no longer known that r is a simple eigenvalue.

Contrast these results with Theorems 4.4 and 4.7.)

Proof

If A is irreducible (1) and (2) follow by Theorem 4.7, whereas (3) follows by Theorem 4.4.

If A is reducible we can transform it into a normal form (4.29).

(1) Since B_{jj} is irreducible

$$r_j = \rho(B_{jj}) \geqslant 0.$$

Then

$$r = \max_j \{r_j\} = \rho(A) \geqslant 0.$$

Note
If $r = 0$, $B_{jj} = [0]$ ($j = 1, 2, \ldots, m$). In this case the normal form of A is strictly triangular.

(2) We make use of a proof based on a continuity argument† since on replacing each zero entry in A by $e_i > 0$ where

$$\lim_{i \to \infty} e_i = 0,$$

we obtain a sequence $\{A_i\}$ of primitive matrices, having the property

$$\lim_{i \to \infty} A_i = A.$$

Since A_i is primitive

$$r_i = \rho(A_i) > 0$$

and the corresponding normalised eigenvector $\mathbf{x}_i > \mathbf{0}$ where

$$A_i \mathbf{x}_i = r_i \mathbf{x}_i \qquad (i = 1, 2, \ldots)$$

From a theorem in set theory it follows that there is a limit point $\mathbf{x}_0 \geqslant \mathbf{0}$ of the sequence $\{x_i\}$ such that

$$A\mathbf{x}_0 = r\mathbf{x}_0$$

where

$$r = \lim_{i \to \infty} r_i.$$

(3) Considering that A is in the form

$$\begin{bmatrix} A_{11} & A_{12} \\ 0 & A_{22} \end{bmatrix}$$

it is clear that r must be an eigenvalue of A_{11} or A_{22}.

† What is stated here is that the eigenvalues $\lambda_1, \lambda_2, \ldots, \lambda_n$ of a matrix $A = [a_{ij}]$ of order $(n \times n)$ depend *continuously* on the elements $a_{11}, a_{12}, \ldots, a_{nn}$.
The eigenvalues $\lambda_1, \lambda_2, \ldots, \lambda_n$ are the roots of the characteristic equation

$$C(\lambda) = |\lambda I - A| = 0.$$

Since the coefficients of the polynomial $C(\lambda)$ depend continuously on the elements a_{ij} ($i, j = 1, 2, \ldots, n$) is is only necessary to prove that roots of the polynomial depend continuously on the coefficients. This is a known result in the theory of polynomials and is of no direct concern in the theory of matrices. (For further details see [B5] p. 191.)

It will be assumed that A_{11} and A_{22} are irreducible. (If this is not the case we consider A in its normal form and apply the following argument to the B_{jj}.)

If any entry of either A_{11} or A_{22} is increased, then r is increased (see Theorem 4.4).

On the other hand, an increase in an element of A_{12} does not affect r.

Example 4.13

Consider the matrix

$$A = \begin{bmatrix} 2 & 1 & 0 & 0 & 2 \\ 0 & 0 & 4 & 0 & 0 \\ 0 & 0 & 0 & 1 & 0 \\ 0 & 2 & 0 & 0 & 0 \\ 1 & 0 & 0 & 0 & 0 \end{bmatrix}.$$

(1) Find a matrix P such that PAP' is in a normal form.
(2) What is the value of r?

Solution

(1) With

$$\sigma = \begin{pmatrix} 1 & 2 & 3 & 4 & 5 \\ 5 & 1 & 2 & 3 & 4 \end{pmatrix},$$

the corresponding permutation matrix P is

$$P = \begin{bmatrix} 0 & 0 & 0 & 0 & 1 \\ 1 & 0 & 0 & 0 & 0 \\ 0 & 1 & 0 & 0 & 0 \\ 0 & 0 & 1 & 0 & 0 \\ 0 & 0 & 0 & 1 & 0 \end{bmatrix}$$

and

$$PAP' = \left[\begin{array}{cc|ccc} 0 & 1 & 0 & 0 & 0 \\ 2 & 2 & 1 & 0 & 0 \\ \hline 0 & 0 & 0 & 4 & 0 \\ 0 & 0 & 0 & 0 & 1 \\ 0 & 0 & 2 & 0 & 0 \end{array}\right].$$

This is a normal form of A since both

$$A_{11} = \begin{bmatrix} 0 & 1 \\ 2 & 2 \end{bmatrix} \quad \text{and} \quad A_{22} = \begin{bmatrix} 0 & 4 & 0 \\ 0 & 0 & 1 \\ 2 & 0 & 0 \end{bmatrix}$$

are irreducible.

Note
Another normal form is obtained by taking

$$\sigma = \begin{pmatrix} 1 & 2 & 3 & 4 & 5 \\ 5 & 1 & 2 & 3 & 4 \end{pmatrix}.$$

In fact a normal form of A is unique up to permutations by blocks.

(2) $\mathrm{r}(A_{11}) = 2.73, \mathrm{r}(A_{22}) = 2$

hence

$$\mathrm{r}(A) = 2.73. \quad \blacksquare$$

It is not surprising that many properties of irreducible matrices also apply, with some modification, to reducible matrices.

For example it was shown in Lemma 4.1 that when A is irreducible

$$B(r) > 0.$$

Since for a reducible matrix, corresponding to the eigenvalue r, the eigenvector $\mathbf{x} \geqslant \mathbf{0}$ it follows that

$$B(r) \geqslant 0.$$

We now prove a generalisation of the above result applicable to all nonnegative matrices.

Theorem 4.10
Given $A(n \times n) = [a_{ij}]$, a nonnegative matrix with $\rho(A) = r$, then

$$B(\lambda) = \text{adj } [\lambda I - A] \geqslant 0 \text{ for } \lambda \geqslant r.$$

Proof
We make use of a proof by induction.

For $n = 1$, the result is obviously correct.

We assume that for $\lambda \geqslant \lambda_A$ (λ_A depending on the matrix A)

$$B(\lambda) \geqslant 0 \quad \text{and} \quad \frac{\mathrm{d}B}{\mathrm{d}\lambda} \geqslant 0$$

for a nonnegative matrix A of order $(n-1) \times (n-1)$.

Let

$$B(\lambda) = [B_{ij}]$$

B_{ji} or more correctly $B_{ji}(\lambda)$ is the cofactor of the (i, j)th entry in $[\lambda I - A]$.

In what follows we use the following convention: we associate the index set $N = \{1, 2, \ldots, n\}$ with the rows and columns of $[\lambda I - A]$. We also associate N with matrices (and determinants) obtained by deleting rows and columns from $[\lambda I - A]$. For example B_{ij} is obtained by deleting the jth row and the ith column from $|\lambda I - A|$. So the $(n-1)$ columns of B_{ij} can be indexed by the set $\{j \in N; j \neq i\}$.

On expanding $|\lambda I - A|$ by its kth row, we obtain

$$\begin{aligned} C(\lambda) = &-a_{k1}B_{1k} - a_{k2}B_{2k} - \ldots \\ &+ (\lambda - a_{kk})B_{kk} - \ldots - a_{kn}B_{nk} \end{aligned} \tag{4.30}$$

We now construct a new matrix

$$B_k = [B_{ij}^{(k)}] \qquad (k \in N)$$

of order

$$(n-1) \times (n-1)$$

by deleting the kth row and the kth column from B.

Notice that each

$$B_{jk} \ (j \in N;\ j \neq k)$$

has a column all of whose $(n-1)$ entries are the entries of the kth column of $[\lambda I - A]$, (the entry not included is $(\lambda - a_{kk})$).

On expanding B_{jk} by this column, we obtain

$$B_{jk} = -a_{1k}B_{j1}^{(k)} - a_{2k}B_{j2}^{(k)} - \ldots - a_{nk}B_{jn}^{(k)} \qquad (j \in N;\ j \neq k) \tag{4.31}$$

From (4.30) and (4.31) we obtain

$$C(\lambda) = (\lambda - a_{kk})B_{kk} - \sum_{i,j=1}^{n} a_{ik}a_{kj}B_{ji}^{(k)} \qquad (i, j \neq k) \tag{4.32}$$

$B_{kk}(\lambda)$ is a polynomial of degree $(n-1)$ in λ.

Let λ_k be the largest nonnegative root of B_{kk}.

Then by the above hypothesis

$$B_k(\lambda) \geqslant 0 \ \text{ for } \ \lambda \geqslant \lambda_k$$

so that on setting $\lambda = \lambda_k$ in (4.32), we find

$$C'(\lambda_k) \leqslant 0.$$

But, as we have seen in Lemma 4.1

$$C(\lambda) > 0 \text{ for } \lambda > r$$

from which it follows that

$$r \geqslant \lambda_k \qquad (k \in N)$$

so, for example,

$$B_k(\lambda) \geqslant 0 \text{ for } \lambda \geqslant r.$$

We can now prove that $dB/d\lambda \geqslant 0$ for $\lambda \geqslant r$.

$dB_{ij}/d\lambda$ is the sum of $(n-1)$ determinants, the kth one being identical to B_{ij} except for the entries in the kth column which are the derivatives of the corresponding entries in B_{ij} (see Section 1.6). The only non-zero entry in this column results from differentiating $(\lambda - a_{kk})$. Every column of B_{ij} has such an entry except for the jth (it has been deleted when forming B_{ij} from $|\lambda I - A|$).

Expanding B_{ij} by this kth column, the result is $B_{ij}^{(k)}$.

It follows that

$$\frac{dB_{ij}}{d\lambda} = \sum_{k=1}^{n} B_{ij}^{(k)} \geqslant 0 \qquad (k \neq i,\ k \neq j)$$

so that

$$\frac{dB}{d\lambda} \geqslant 0 \text{ for } \lambda \geqslant r$$

But $B(r) \geqslant 0$, hence

$$B(\lambda) \geqslant 0 \text{ for } \lambda \geqslant r.$$

Note

If A is an irreducible matrix it is not difficult to amend the above proof and show that

$$B(\lambda) = \text{adj}\,(\lambda I - A) > 0 \text{ for } \lambda \geqslant r.$$

Example 4.14

Given an irreducible matrix $A \geqslant 0$, with $\rho(A) = r$, show that

$$[\lambda I - A]^{-1} > 0 \text{ for } \lambda > r.$$

(See also Section 1.14.)

Solution

In Lemma 4.1 we have seen that

$$C(\lambda) = |\lambda I - A| > 0 \text{ for } \lambda > r.$$

Also by the above

$$\text{adj } [\lambda I - A] > 0 \text{ for } \lambda > r$$

the result follows since

$$[\lambda I - A]^{-1} = \frac{1}{C(\lambda)} \text{adj } [\lambda I - A].$$

PROBLEMS

4.1 Consider

$$(A\mathbf{x})_j = \sum_r a_{jr} x_r$$

for

$$A = \begin{bmatrix} 2 & 2 & 1 \\ 1 & 3 & 1 \\ 1 & 2 & 2 \end{bmatrix}$$

and

$$\mathbf{x} \in S = \{\mathbf{x}' = (x_1, x_2, x_3), x_i \geqslant 0\}$$

(1) $\mathbf{x} = \begin{bmatrix} 1 \\ 2 \\ 1 \end{bmatrix}$ (2) $\mathbf{x} = \begin{bmatrix} 1 \\ 2 \\ 3 \end{bmatrix}$, (3) $\mathbf{x} = \begin{bmatrix} 1 \\ 0 \\ 1 \end{bmatrix}$ (4) $\mathbf{x} = \begin{bmatrix} 1 \\ 2 \\ 0 \end{bmatrix}$ and

(5) $\mathbf{x} = \begin{bmatrix} 1 \\ 1 \\ 1 \end{bmatrix}$.

For each $\mathbf{x}$ evaluate

$$p(\mathbf{x}) = \min_j \frac{(A\mathbf{x})_j}{x_j},$$

excluding $x_j = 0$ and

$$q(\mathbf{x}) = \max_j \frac{(A\mathbf{x})_j}{x_j},$$

defining $q(\mathbf{x}) = \infty$ for $x_j = 0$.

Show that (for the given vectors in this example, but generally for all $\mathbf{x} \in S$)

$$\max_{\mathbf{x}\in S} p(\mathbf{x}) = \min_{\mathbf{x}\in S} q(\mathbf{x})$$

and find this value (equal to the spectral radius of A).

4.2 (1) Consider the matrices

$$B = \begin{bmatrix} 9 & 4 & 3 \\ 1 & 12 & 3 \\ 5 & 4 & 7 \end{bmatrix} \quad \text{and} \quad A = \begin{bmatrix} 9 & 4 & 3 \\ 2 & 12 & 3 \\ 5 & 4 & 7 \end{bmatrix}$$

so that $A \geqslant B > 0$. Verify Theorem 4.2 that

$$\rho(A) > \rho(B).$$

(2) Verify that the eigenvector corresponding to $\rho(B)$ is positive.

4.3 For both the matrix (1)A and (2)B in Problem 4.2 show that (Theorem 4.5).

$$\min_i \sum_{j=i}^{n} a_{ij} \leqslant r \leqslant \max_i \sum_{j=1}^{n} a_{ij}$$

where r is the spectral radius of the matrix.

4.4 $A \geqslant 0$ is an irreducible matrix with $r = \rho(A)$ and a corresponding eigenvector $\mathbf{x} > \mathbf{0}$. Given that λ is another eigenvalue of A and $\mathbf{y}$ the corresponding eigenvector, show that it is not possible that

$$\mathbf{y} \geqslant \mathbf{0}.$$

Note This implies that A cannot have two (linearly independent) nonnegative eigenvectors.

4.5 Classify the following matrices into reducible and irreducible classes.

$$(1) \begin{bmatrix} 0 & 1 & 1 \\ 0 & 1 & 0 \\ 1 & 0 & 1 \end{bmatrix} \quad (2) \begin{bmatrix} 0 & 0 & 1 & 1 \\ 0 & 0 & 1 & 0 \\ 1 & 1 & 0 & 0 \\ 1 & 0 & 0 & 0 \end{bmatrix} \quad (3) \begin{bmatrix} 1 & 1 & 1 & 1 \\ 0 & 1 & 0 & 1 \\ 1 & 0 & 1 & 0 \\ 0 & 0 & 1 & 1 \end{bmatrix}$$

and

$$(4) \begin{bmatrix} 0 & 0 & 1 & 0 \\ 0 & 1 & 0 & 1 \\ 0 & 1 & 0 & 0 \\ 1 & 0 & 0 & 0 \end{bmatrix}$$

4.6 Given the irreducible matrix

$$A = \begin{bmatrix} 0 & 1 & 0 & 1 & 0 \\ 0 & 0 & 1 & 0 & 0 \\ 1 & 0 & 0 & 0 & 0 \\ 0 & 0 & 0 & 0 & 1 \\ 0 & 0 & 1 & 0 & 0 \end{bmatrix}$$

and

$$\mathbf{y} = [1 \;\; 0 \;\; 0 \;\; 0 \;\; 0]'$$

Verify that (Theorem 4.6)

(1) $\mathbf{z} = [A + I]\mathbf{y}$ has more non-zero elements than $\mathbf{y}$, and
(2) $[I + A]^4 > 0$.

4.7 Consider the cyclic matrix (see [B22])

$$A = \begin{array}{c} \\ 1 \\ 2 \\ 3 \\ 4 \\ 5 \\ 5 \end{array} \begin{array}{c} \begin{array}{cccccc} 1 & 2 & 3 & 4 & 5 & 6 \end{array} \\ \begin{bmatrix} 0 & 1 & 0 & 0 & 0 & 0 \\ 0 & 0 & 1 & 1 & 0 & 1 \\ 1 & 0 & 0 & 0 & 1 & 0 \\ 1 & 0 & 0 & 0 & 0 & 0 \\ 0 & 1 & 0 & 0 & 0 & 0 \\ 0 & 0 & 0 & 0 & 1 & 0 \end{bmatrix} \end{array} .$$

(1) Find the period of A.
(2) Determine the subclasses C_r of the index set $N = \{1, 2, 3, 4, 5, 6\}$ associated with A.
(3) Find the permutation matrix P such that the indices of the matrix

$$PAP^{-1}$$

are in a sequence defined by $\{C_1, C_2, C_0\}$.

(4) $$PAP^{-1} = \begin{bmatrix} 0 & 0 & A_3 \\ A_1 & 0 & 0 \\ 0 & A_2 & 0 \end{bmatrix} .$$

Determine A_1, A_2 and A_3.

(5) $PA^{d}P^{-1} = \text{diag}\{Q_3, Q_2, Q_1\}$. Evaluate Q_3, Q_2 and Q_1.

4.8 Evaluate the spectral radius of the matrix A in Problem 4.7.

4.9 Consider the matrix

$$A = \begin{bmatrix} 1 & 0 & 2 & 0 \\ 0 & 1 & 0 & 2 \\ 1 & 1 & 0 & 0 \\ 0 & 1 & 0 & 1 \end{bmatrix}.$$

(1) Show that A is reducible by finding the permutation matrix P by such that

$$PAP' = \begin{bmatrix} B & 0 \\ C & D \end{bmatrix}$$

where B and D are square matrices.

(2) Evaluate $\rho(A)$.

(3) Evaluate $\rho(A_{rr})$ ($r = 1, 2, 3, 4$) where A_{rr} are the principal minors of A.

Note that

$$\rho(A) = \rho(A_{rr})$$

for at least one r($r = 1, 2, 3, 4$).

5

M-Matrices

5.1 INTRODUCTION

In various considerations of convergence of numerical techniques for solving both linear and non-linear systems of equations, in economic theory and many other applications, the structure of the matrix A defining the problem takes the following form:

$$A = \begin{bmatrix} a_{11} & -a_{12} & -a_{13} & \dots & -a_{1n} \\ -a_{21} & a_{22} & -a_{23} & \dots & -a_{2n} \\ \vdots & & & & \vdots \\ -a_{n1} & -a_{n2} & -a_{n3} & \dots & a_{nn} \end{bmatrix}.$$

The entries a_{ij} are all nonnegative. So A can be written as

$$A = sI - B \tag{5.1}$$

where $s > 0$, $B \geqslant 0$ and $s \geqslant \rho(B)$.

For example

$$A = \begin{bmatrix} 3 & -1 & -2 \\ -4 & 2 & -1 \\ -5 & -2 & 5 \end{bmatrix} = 5\begin{bmatrix} 1 & 0 & 0 \\ 0 & 1 & 0 \\ 0 & 0 & 1 \end{bmatrix} - \begin{bmatrix} 2 & 1 & 2 \\ 4 & 3 & 1 \\ 5 & 2 & 0 \end{bmatrix}$$

so that

$$B = \begin{bmatrix} 2 & 1 & 2 \\ 4 & 3 & 1 \\ 5 & 2 & 0 \end{bmatrix}$$

and

$$s = 5.$$

But s is not unique, indeed we can choose any value for $s \geqslant 5$.

For example let $s = 7$, then

$$A = 7\begin{bmatrix} 1 & 0 & 0 \\ 0 & 1 & 0 \\ 0 & 0 & 1 \end{bmatrix} - \begin{bmatrix} 4 & 1 & 2 \\ 4 & 5 & 1 \\ 5 & 2 & 2 \end{bmatrix}.$$

Since B is a nonnegative matrix, the importance and application of Chapter 4 is obvious. Matrices having the structure (5.1) are called M-matrices (Fiedler and Ptak [8] and [9] call them K-matrices).

If $S > \rho B)$ in (5.1) then M-matrix is (obviously non-singular. Following the established literature we use the notation:

$$Z = \{A = [a_{ij}];\ a_{ij} \leqslant 0,\ i \neq j\}.$$

We shall introduce the notation for non-singular M-matrices as:

$$\mathscr{M} = \{A : A \in Z;\ A = sI - B;\ B \geqslant 0;\ s > 0;\ s > \rho(B)\}.$$

We shall use the notation $\mathscr{M}_0$ to include M-matrices which are singular, that is M-matrices for which $S = \rho(B)$.

Notice that if $A \in M$, we can always find $t > 0$ such that

$$I - tA \geqslant 0.$$

For example, for the above matrix A, choose $t = 1/5$, then

$$\begin{bmatrix} 1 & 0 & 0 \\ 0 & 1 & 0 \\ 0 & 0 & 1 \end{bmatrix} - 1/5\begin{bmatrix} 3 & -1 & -2 \\ -4 & 2 & -1 \\ -5 & -2 & 5 \end{bmatrix} = \begin{bmatrix} 2/5 & 1/5 & 2/5 \\ 4/5 & 3/5 & 1/5 \\ 1 & 2/5 & 0 \end{bmatrix}.$$

Another implication of this definition, is that the diagonal entries are all positive. In consideration of general M-matrices, including singular M-matrices, this restriction is lifted.

5.2 NON-SINGULAR M-MATRICES

In this section we shall demonstrate various properties of M-matrices.

Lemma 5.1

If $A \in \mathscr{M}$, then A is non-singular and $A^{-1} \geqslant 0$.

Proof
A is obviously non-singular since $S > \rho(B)$. We can write

$$A = sI - B = s\left[I - \frac{B}{s}\right],$$

then

$$A^{-1} = \frac{1}{s}\left[I - \frac{B}{s}\right]^{-1}$$

$$= \frac{1}{s}\left[I + \frac{B}{s} + \frac{B^2}{s^2} + \ldots\right]$$

(since $\rho(B/s) < 1$, see Section 1.14).
But

$$B \geqslant 0 \quad \text{and} \quad s > 0,$$

so that

$$\left[\frac{B}{s}\right]^k \geqslant 0 \qquad (k = 1, 2, \ldots)$$

hence

$$A^{-1} \geqslant 0.$$

Note
If B is irreducible, then $A^{-1} > 0$, (see Corollary 2 of Lemma 4.2).

Corollary
If $A^{-1} \geqslant 0$, then $S > \rho(B)$.

Proof
Since B is non-negative, $\rho(B)$ is an eigenvalue of B and let $\mathbf{x} \geqslant \mathbf{0}$ be a corresponding eigenvector, so that

$$B\mathbf{x} = \rho(B)\mathbf{x}$$

and

$$[sI - B]\,\mathbf{x} = (s - \rho(B))\mathbf{x}.$$

Since A is non-singular,

$$s - \rho(B) \neq 0.$$

Multiplying the above equation by $(s - \rho(B))^{-1}\,[sI - B]^{-1}$, we obtain

$$(s - \rho(B))^{-1}\mathbf{x} = [sI - B]^{-1}\mathbf{x}.$$

Since

$$[sI - B]^{-1} \geqslant 0 \quad \text{and} \quad \mathbf{x} \geqslant \mathbf{0},$$

then

$$(s - \rho(B))^{-1} > 0$$

so that

$$s > \rho(B).$$

Definition 5.1
Given a matrix $A = [a_{ij}]$ and a positive diagonal matrix

$$D = \text{diag}\{d_1, d_2, \ldots, d_n\}$$

then AD is called *strictly diagonally dominant* (s.d.d.) if

$$|a_{ii}|d_i > \sum_{i \neq j} |a_{ij}|d_j \qquad (i = 1, 2, \ldots, n). \tag{5.2}$$

If these inequalities hold A is said to have *strictly dominant diagonal* (*s.d.d.*). or just a dominant diagonal.

If instead of the inequality above, equality is valid for some i (but not all i), then AD is called *quasi-diagonally dominant* (*q.d.d.*).

Lemma 5.2
If $A \in \mathcal{M}$ then there exists a diagonal matrix D such that AD is s.d.d.

Proof
By Lemma 5.1

$$A^{-1} \geqslant 0.$$

Let $\mathbf{e}$ be the vector having all entries unity. Then

$$A^{-1}\mathbf{e} = \mathbf{d} = [d_1, d_2, \ldots, d_n]' > \mathbf{0}.$$

Let

$$D = \text{diag}\{d_1, d_2, \ldots, d_n\}$$

then

$$AD\mathbf{e} = A\mathbf{d} = \mathbf{e} > \mathbf{0}.$$

In terms of components, the above inequality is written as:

$$-a_{i1}d_1 - a_{i2}d_2 - \ldots + a_{ii}d_i - \ldots - a_{in}d_n > 0$$

so that

$$a_{ii}d_i > \sum_{i \neq j} a_{ij}d_j. \qquad (i = 1, 2, \ldots, n).$$

Let

$$AD = C = [c_{ij}].$$

Then the last set of inequalities is written as

$$c_{ii} > \sum_{i \neq j} c_{ij} \qquad (i = 1, 2, \ldots, n).$$

Various lemmas in this chapter actually characterise M-matrices and so are equivalent. For example, it can be shown that if a matrix A has s.d.d. (or indeed q.d.d.) then A is non-singular.

Lemma 5.3
If a matrix $A(n \times n)$ has s.d.d., then A is non-singular.

Proof
Assume the contrary, that A is singular. Then $AD = C$ where $C = [c_{ij}]$ and $D = \text{diag}\,\{d_1, d_2, \ldots, d_n\}$, $d_i >$ (all i) is also singular.

So [B7 p. 103] there is a vector

$$\mathbf{x} = [x_1, x_2, \ldots, x_n]' \neq \mathbf{0}$$

such that

$$\mathbf{x}'C = \mathbf{0}$$

or

$$\sum_{i=1}^{n} x_i c_{ij} = 0 \qquad (j = 1, 2, \ldots, n) \tag{5.3}$$

Let K be an index set such that

$$|x_j| \geqslant |x_i| \text{ whenever } j \in K$$

and

$$|x_j| > |x_i| \text{ whenever } j \in K$$

but

$$i \notin K \qquad (i = 1, 2, \ldots, n).$$

For example, if

$$\mathbf{x} = [2 \quad 1 \quad -1 \quad -2]'$$

then

$$K = \{1, \quad 4\}.$$

Choosing some $j \in K$, we can write the relevant equations in (5.3) as

$$x_j c_{jj} + \sum_{i \neq j} x_i c_{ij} = 0$$

so that

$$|x_j|\,|c_{jj}| \leqslant \sum_{\substack{i \in K \\ i \neq j}} |x_i|\,|c_{ij}| + \sum_{i \notin K} |x_i|\,|c_{ij}| \leqslant \sum_{i \neq j} |x_j|\,|c_{ij}| \tag{5.4}$$

But A has s.d.d., so that

$$|c_{jj}| > \sum_{\substack{i=1 \\ i \neq j}}^{n} |c_{ij}|$$

From which it follows that

$$|x_j|\,|c_{jj}| > \sum_{\substack{i=1 \\ i \neq j}}^{n} |x_j|\,|c_{ij}| \tag{5.5}$$

(5.5) is a contradiction of (5.4).

Hence no such $\mathbf{x}$ exists, and C is non-singular, so A is non-singular.

Lemma 5.4

$A(n \times n) \in \mathscr{M}$ has all its principal minors $|A(\alpha)|$ positive. (For notation see Section 1.3.).

Proof

The principal minor of A of order n is $|A|$.

$$|A| = |sI - B| > 0 \qquad \text{(see Lemma 4.1).}$$

Let

$$\alpha = \{1, 2, \ldots, k-1,\ k+1, \ldots, n\},$$

then

$$A(\alpha) = sI - B_1$$

where I is the unit matrix of order $(n-1) \times (n-1)$ and B_1 is obtained by deleting the kth row and the kth column from B.

By Theorem 4.2

$$\rho(B_1) \leqslant \rho(B).$$

Since B_1 is a nonnegative matrix

$$|sI - B_1| > 0$$

for

$$s > \rho(B_1).$$

Hence

$$|A(\alpha)| = |sI - B_1| > 0$$

since

$$s > \rho(B) \geqslant \rho(B_1).$$

The above result is valid for $k = 1, 2, \ldots, n$, that is for all principal minors of A of order $(n-1)$. By similar argument it can be shown that the above conclusion is true for all principal minors of order $n > 1$.

For $n = 1$

$$|A| = a_{11} > 0.$$

Corollary
If $A \in \mathcal{M}$ and A is symmetric then A is positive definite.

Proof
See Theorem 3.4 and Lemma 5.4.

Lemma 5.5
Consider $A \in \mathcal{M}$.

(1) Every real eigenvalue of A is positive.
(2) The real part of each eigenvalue of A is positive.

Proof
(1) $A = SI - B; s > 0$.
Let $\mathbf{x}$ be an eigenvector of B corresponding to the eigenvalue $\rho(B) > 0$, then

$$B\mathbf{x} = \rho(B)\mathbf{x}$$

so that

$$A\mathbf{x} = [sI - B]\,\mathbf{x} = [s - \rho(B)]\,\mathbf{x}.$$

Hence

$$s - \rho(B)$$

is a (real) eigenvalue of A.
Since $A \in \mathcal{M}$,

$$s > \rho(B),$$

and the proof follows.
(2) Let $k > 0$ be such that

$$kI - A \geqslant 0.$$

Let $\lambda = \alpha + i\beta$ be *any* eigenvalue of A, then $(k - \lambda)$ is an eigenvalue of $[kI - A]$, so that

$$|k - \lambda| \leqslant \rho[kI - A].$$

There exists a real eigenvalue λ_0 (say) of A such that

$$k - \lambda_0 = \rho[kI - A].$$

By (1) above

$$\lambda_0 > 0$$

so that

$$|k - \lambda| \leqslant k - \lambda_0 < k$$

or

$$[(k - \alpha)^2 + \beta^2]^{1/2} < k$$

which is possible iff

$$\alpha > 0.$$

This proves part (2) of the lemma. ∎

The sign of the real parts of eigenvalues of a matrix A is of the greatest importance when considering the stability of a system.

The matrix A is said to be *negative* (positive) *stable* if all eigenvalues of A have negative (positive) real parts.

Lyapunov's theorem (see B8) states that all eigenvalues of A have negative (positive) real parts iff there exists a symmetric positive definite matrix P such that

$$AP + PA' = -Q \quad \text{(say)}$$

and Q is positive (negative) definite.

Under certain conditions the matrix P can be chosen to be positive diagonal D (say). The following statements are equivalent for a matrix A.

(1) There exists a positive diagonal matrix D such that

$$AD + DA'$$

is positive definite.

(2) There exists a positive diagonal matrix D such that

$$\mathbf{x}'AD\mathbf{x} > 0$$

for all

$$\mathbf{x} \neq \mathbf{0}.$$

To illustrate the equivalence in the case of a matrix A of order (2×2), let

$$A = \begin{bmatrix} a_{11} & a_{12} \\ a_{21} & a_{22} \end{bmatrix}, \quad D = \text{diag}\{d_1, d_2\} \quad \text{and} \quad \mathbf{x} = [x_1, x_2]'.$$

Then

$$AD + DA' = \begin{bmatrix} 2a_{11}d_1 & a_{12}d_2 + a_{21}d_1 \\ a_{12}d_2 + a_{21}\,d_1 & 2a_{22}d_2 \end{bmatrix}$$

is positive definite iff

(a) $a_{11}d_1 > 0, a_{22}d_2 > 0,$ and

(b) $4a_{11}a_{22}d_1d_2 > (a_{12}d_2 + a_{21}d_1)^2$.

Also

$$\mathbf{x}'AD\mathbf{x} = a_{11}d_1x_1^2 + a_{22}d_2x_2^2 + (a_{12}d_2 + a_{21}d_1)x_1x_2.$$

So

$$\mathbf{x}'AD\mathbf{x} > 0 \text{ iff}$$

$$a_{11}d_1y^2 + (a_{12}d_2 + a_{21}d_1)y + a_{22}d_2 > 0$$

where

$$y = \frac{x_1}{x_2}.$$

Given that $a_{11}d_1 > 0$ and $a_{22}d_2 > 0$, the quadratic is positive for all y, if its roots are complex that is if

$$4a_{11}a_{22}d_1d_2 > (a_{12}d_2 + a_{21}d_1)^2$$

this proves the equivalence of (1) and (2) when A is of order (2×2). The equivalence is of course valid for a matrix A of order $(n \times n)$.

In Lemma 5.8 it is shown if $A \in \mathcal{M}$ then

$$\mathbf{x}'AD\mathbf{x} > 0 \text{ for } \mathbf{x} \neq \mathbf{0}.$$

Lemma 5.6
If $A, B \in \mathcal{M}$ and $B \geqslant A$, then

(a) $A^{-1} \geqslant B^{-1}$

(b) $|B| \geqslant |A| > 0$.

Proof
(1) $\exists t > 0$ such that

$$P = I - tA \geqslant I - tB = Q \geqslant 0.$$

Since $\rho(Q)$ is an eigenvalue of Q

$$|Q - \rho(Q)I| = 0$$

or

$$|I - tB - \rho(Q)I| = |(1 - \rho(Q))I - tB| = 0.$$

So

$$(1 - \rho(Q))$$

is a real eigenvalue of $tB \in \mathcal{M}$.

By Lemma 5.5

$$1 - \rho(Q) > 0$$

so that

$$0 \leqslant \rho(Q) < 1$$

From Section 1.14

$$I + Q + Q^2 + \ldots = [I - Q]^{-1}$$
$$= [tB]^{-1}.$$

Similarly

$$I + P + P^2 + \ldots = [tA]^{-1}.$$

Since $P \geqslant Q \geqslant 0$, then

$$P^k \geqslant Q^k \geqslant 0 \qquad (k = 1, 2, \ldots).$$

It follows that

$$[tA]^{-1} \geqslant [tB]^{-1} \geqslant 0.$$

But $t > 0$, hence

$$A^{-1} \geqslant B^{-1}.$$

(2) If $n = 1$

$$|B| = b_{11} \geqslant a_{11} = |A| > 0.$$

Let $n > 1, N = 1, 2, \ldots, n$ and $\alpha \subset N$ (for notation see Section 1.3).

Assume that

$$|B(\alpha)| \geqslant |A(\alpha)| > 0.$$

Notice that $A(\alpha), B(\alpha) \in \mathcal{M}$ and $B(\alpha) \geqslant A(\alpha)$.

Let $\alpha = \{1, 2, \ldots, n-1\}$ and denote

$$A(\alpha) = A_1 \quad \text{and} \quad B(\alpha) = B_1.$$

Then

$$A(n \times n) = \left[\begin{array}{ccc:c} & & & x \\ & & & x \\ A_1 & & & \vdots \\ \hdashline xx & \dots & & a_{nn} \end{array}\right] \quad \text{and} \quad B(n \times n) = \left[\begin{array}{ccc:c} & & & x \\ & & & x \\ B_1 & & & \vdots \\ \hdashline xx & \dots & & b_{nn} \end{array}\right]$$

Since the cofactor of a_{nn} is $|A_1|$ and the cofactor of b_{nn} is $|B_1|$ the form of the matrices A^{-1} and B^{-1} is as follows:

$$A^{-1} = \frac{1}{|A|}\begin{bmatrix} xx & \dots & x \\ xx & \dots & x \\ \vdots & & \vdots \\ xx & \dots & |A_1| \end{bmatrix} \quad \text{and} \quad B^{-1} = \frac{1}{|B|}\begin{bmatrix} xx & \dots & x \\ xx & & x \\ \vdots & & \vdots \\ xx & \dots & |B_1| \end{bmatrix}$$

But by (1) above

$$A^{-1} \geqslant B^{-1} \geqslant 0,$$

hence

$$\frac{|A_1|}{|A|} \geqslant \frac{|B_1|}{|B|} \geqslant 0$$

so that

$$|A| > 0 \quad \text{and} \quad |B| > 0.$$

It follows that

$$|B| \geqslant \frac{|A|}{|A_1|}\,|B_1| \geqslant |A| > 0 \qquad (\text{since } |B_1| \geqslant |A_1|). \quad \blacksquare$$

Before proceeding with the next lemma we need a preliminary result. We can write $A(n \times n) \in \mathcal{M}$ in the partitioned form

$$A = \left[\begin{array}{c:c} A_1 & \mathbf{a} \\ \hdashline \mathbf{b}' & a_{nn} \end{array}\right]$$

where A_1 is a principal submatrix defined above, $\mathbf{a}$ and $\mathbf{b}$ are vectors of order $(n-1)$.

The form of A^{-1}, showing its nth row is

$$A^{-1} = \frac{1}{|A|}\begin{bmatrix} xx & \dots & x \\ \vdots & & \\ xx & \dots & x \\ \mathbf{c}' & & |A_1| \end{bmatrix}$$

where $\mathbf{c}$ is a vector of order $(n-1)$.

The (zero) vector of the first $(n-1)$ entries of the nth row of $[A^{-1}A]$ is

$$\frac{1}{|A|}(\mathbf{c}'A_1\ |A_1|\mathbf{b}') = \mathbf{0}'$$

so that

$$\mathbf{c}' = -|A_1|\mathbf{b}'A_1^{-1}.$$

The nth entry is

$$\frac{1}{|A|}(\mathbf{c}'\mathbf{a} + |A_1|a_{nn}) = 1.$$

Substituting for $\mathbf{c}'$, we finally obtain

$$a_{nn} - \mathbf{b}'A_1^{-1}\mathbf{a} = \frac{|A|}{|A_1|} > 0 \quad \text{(by Lemma 5.6).} \qquad (5.6)$$

Lemma 5.7

$A(n \times n) \in \mathcal{M}$ can be written as the product

$$A = LU$$

where L and U are lower and upper triangular matrices respectively such that

(1) all diagonal entries are positive
(2) all other entries are non-positive.

Note (1) and (2) imply that L and U are themselves M-matrices.

Proof

(1) We use induction.

For $n = 1$, the result is trivial.

Assume the result is true for all matrices $\in \mathcal{M}$ of order $(k \times k)$ where $k < n$.

For example, A_1 (defined above) can be written as

$$A_1 = L_1 U_1$$

where L_1 and U_1 have positive diagonals.

It is now but a simple step to show that A can be written as

$$A = \begin{bmatrix} A_1 & \mathbf{a} \\ \mathbf{b}' & a_{nn} \end{bmatrix} = \begin{bmatrix} L_1 & 0 \\ \mathbf{b}'U_1^{-1} & 1 \end{bmatrix} \begin{bmatrix} U_1 & L_1^{-1}\mathbf{a} \\ 0 & a_{nn} - \mathbf{b}'A_1^{-1}\mathbf{a} \end{bmatrix}.$$

By (5.6) and the above assumption, the diagonal entries of the two matrices on the right-hand side are positive.

(2) Let

$$L = [l_{ij}] \quad \text{and} \quad U = [u_{ij}]$$

so that

$$\begin{bmatrix} a_{11} & -a_{12} & -a_{13} & \dots & -a_{1n} \\ -a_{21} & a_{22} & -a_{23} & \dots & -a_{2n} \\ \vdots & & & & \\ -a_{n1} & -a_{n2} & -a_{n3} & \dots & a_{nn} \end{bmatrix}$$

$$= \begin{bmatrix} l_{11} & 0 & 0 & \dots & 0 \\ l_{21} & l_{22} & 0 & \dots & 0 \\ \vdots & & & & \\ l_{n1} & l_{n2} & l_{n3} & \dots & l_{nn} \end{bmatrix} \begin{bmatrix} u_{11} & u_{12} & u_{13} & \dots & u_{1n} \\ 0 & u_{22} & u_{23} & \dots & u_{2n} \\ \vdots & & & & \\ 0 & 0 & 0 & \dots & u_{nn} \end{bmatrix}$$

where

$$l_{ij} = 0 \quad \text{for} \quad i > j \quad \text{and} \quad l_{ii} > 0$$

and

$$u_{ij} = 0 \quad \text{for} \quad i < j \quad \text{and} \quad u_{ii} > 0.$$

We again use induction, this time on $(i + j)$.

If $i + j = 3$

$$-a_{12} = l_{11}u_{12} \Rightarrow u_{12} \leqslant 0$$

$$-a_{21} = l_{21}u_{11} \Rightarrow l_{21} \leqslant 0$$

since

$$a_{ij} \geqslant 0 \quad (\text{all } i, j; i \neq j).$$

Let $i + j > 3$ but $i \neq j$. Assume that

$$l_{st} \leqslant 0 \quad \text{and} \quad u_{st} \leqslant 0 \qquad (s \neq t)$$

are valid for

$$s + t < i + j.$$

Then for $j > i$

$$-a_{ij} = l_{i1} u_{1j} + l_{i2}u_{2j} + \dots + l_{ii} u_{ij}$$

$$= l_{ii} u_{ij} + \sum_{k<i} l_{ik} u_{kj}.$$

By the above assumption

$$l_{ik} \leqslant 0 \quad \text{for} \quad i + k < i + j$$

and

$$u_{kj} \leqslant 0 \quad \text{for} \quad k+j<i+j$$

so that

$$\sum_{k<i} l_{ik} u_{kj} \geqslant 0.$$

Hence

$$u_{ij} \leqslant 0.$$

If $i>j$ it can be shown similarly that

$$l_{ij} \leqslant 0.$$

Lemma 5.8

Let $N = \{1, 2, \ldots, n\}$

If $A(n \times n) \in \mathscr{M}$,

$$\mathbf{x} = [x_1, x_2, \ldots, x_n]' \neq \mathbf{0}$$

and

$$\mathbf{y} = A\mathbf{x}$$

where

$$\mathbf{y} = [y_1, y_2, \ldots, y_n]',$$

then

(1) $x_k y_k > 0$ for some $k \in N$

(2) There exists a positive diagonal matrix D such that

$$\mathbf{x}'AD\mathbf{x} > 0.$$

Proof

Assume on the contrary, that

$$x_i y_i \leqslant 0 \quad \text{for all} \quad i \in N.$$

Let $\alpha \subseteq N$ be a set of indices such that

$$x_j \neq 0 \quad \text{for} \quad j \in \alpha.$$

Let $A(\alpha)\mathbf{x}(\alpha) = \mathbf{y}(\alpha)$ (for notation see Section 1.3).

By the above assumption $\exists$ a diagonal matrix

$$D = \text{diag}\ \{d_1, d_2, \ldots, d_k\}; \quad d_i \geqslant 0$$

such that

$$\mathbf{y}(\alpha) = D\mathbf{x}(\alpha).$$

It follows that

$$[A(\alpha) + D]\,\mathbf{x}(\alpha) = \mathbf{0}.$$

(1) Since $\mathbf{x}(\alpha) \neq \mathbf{0}$ the above assumption leads us to conclude that

$$[A(\alpha) + D]$$

is singular.

But $A(\alpha) \in \mathcal{M}$, hence

$$[A(\alpha) + D] \in \mathcal{M}$$

and so is non-singular.

The assumption is therefore false and the conclusion of the lemma follows.

(2) Since $x_k y_k > 0$, there exists a number $\epsilon > 0$ such that

$$x_k y_k + \epsilon \sum_{i \neq k} x_i y_i > 0$$

Let

$$D = \text{diag}\,\{d_1, d_2, \ldots, d_n\}$$

where

$$d_i = \epsilon (i \in N, i \neq k)$$

and

$$d_k = 1.$$

The result now follows.

Definition 5.2

If $A \in \mathcal{M}$ and $A = A'$, then A is called a *Stieltjes matrix*.

5.3 REGULAR SPLITTING AND SOLVING SIMULTANEOUS EQUATIONS

Definition 5.3

If $A \in \mathcal{M}$, then A is said to have a (*convergent*) *regular splitting* if

$$A = M - N$$

where M is non-singular and $M^{-1} \geqslant 0$ and $N \geqslant 0$.

To explain the interest of regular splitting we consider one application. A system of linear equations

$$A\mathbf{x} = \mathbf{d}$$

where $A = [a_{ij}]$ is assumed non-singular with positive diagonal entries $a_{ii} > 0$.

One iterative method is to write the equation in the form

$$a_{ii}x_i^{(m+1)} = -\sum_{j=1} a_{ij}x_j^{(m)} + d_i \qquad i = 1, 2, \ldots, n$$

where $x_i^{(0)}$ are the initial estimates.

In matrix form the above can be written as

$$D\mathbf{x}^{(m+1)} = [L + U]\mathbf{x}^{(m)} + \mathbf{d}$$

where L and U are respectively strictly lower and upper triangular matrices whose entries are the negatives of the ones below and above the leading diagonal of A (if $A \in \mathcal{M}$, L and U are both nonnegative).

So

$$A = D - L - U.$$

Since D is non-singular the iterative equation can be written as

$$\mathbf{x}^{(m+1)} = D^{-1}[L + U]\mathbf{x}^{(m)} + D^{-1}\mathbf{d}.$$

This is called the *Jacobi iterative method*.

Rather than having to store both $\mathbf{x}^{(i)}$ and $\mathbf{x}^{(i+1)}$ a modification leads to the *Gauss–Seidel method* which needs the storage of one vector $\mathbf{x}^{(i)}$ only, which is updated at each iteration.

This method uses the iterative equation

$$[D - L]\mathbf{x}^{(m+1)} = U\mathbf{x}^{(m)} + \mathbf{d}$$

or, when $[D - L]$ is non-singular

$$\mathbf{x}^{(m+1)} = [D - L]^{-1}U\mathbf{x}^{(m)} + [D - L]^{-1}\mathbf{d}.$$

If $A = M - N$, both the methods can be written as

$$\mathbf{x}^{(m+1)} = M^{-1}N\mathbf{x}^{(m)} + M^{-1}\mathbf{d} \tag{5.7}$$

where

$$M = D \quad \text{and} \quad N = L + U$$

for the Jacobi method, and

$$M = D - L \quad \text{and} \quad N = U$$

for the Gauss–Seidel method.

Now that the interest in the Definition 5.3 can be appreciated, we must show that the conditions guarantee the convergence of the iteration matrix $M^{-1}N$ in (5.7). From Section 1.14 we know that this necessitates showing that

$$\rho(M^{-1}N) < 1.$$

Lemma 5.9

If $A \in \mathcal{M}$ and $A = M - N$ is a regular splitting then

$$\rho(M^{-1}N) < 1.$$

Proof
Since $A = M - N$

$$M^{-1}N = [A + N]^{-1}N = \{A[I + A^{-1}N]\}^{-1}N$$
$$= [I + H]^{-1}H$$

where

$$H = A^{-1}N.$$

We first show that any eigenvector of $M^{-1}N$ is an eigenvector of H and conversely, and find the relation between the eigenvalues of these two matrices.

Let $\mathbf{x}$ be any eigenvector of H, so that

$$H\mathbf{x} = \lambda\mathbf{x}$$

where λ is the corresponding eigenvalue.

Then

$$[I + H]^{-1}H\mathbf{x} = (1 + \lambda)^{-1}\lambda\mathbf{x}$$

or

$$[M^{-1}N]\,\mathbf{x} = \mu\mathbf{x}$$

where

$$\mu = \frac{\lambda}{1 + \lambda}. \tag{5.8}$$

So that $\mathbf{x}$ is also an eigenvector of $M^{-1}N$ and the corresponding eigenvalue is μ.

Conversely, if μ is any eigenvalue of $M^{-1}N$, then

$$M^{-1}N\mathbf{y} = \mu\mathbf{y}.$$

where $\mathbf{y}$ is the corresponding eigenvector.

So that

$$[I + H]^{-1}H\mathbf{y} = \mu\mathbf{y}$$

or

$$H\mathbf{y} = \mu[I + H]\,\mathbf{y}.$$

Assume that $\mu = 1$, then the above equation can only be true if $\mathbf{y} = \mathbf{0}$. But since $\mathbf{y}$ is an eigenvector, $\mathbf{y} \neq \mathbf{0}$, hence

$$\mu \neq 1.$$

So we can write the above as

$$H\mathbf{y} = \lambda\mathbf{y}$$

where

$$\lambda = \frac{\mu}{1-\mu}. \tag{5.9}$$

So $\mathbf{y}$ is also an eigenvector of H and the corresponding eigenvalue is λ. Notice that (5.8) and (5.9) are equivalent.

Since $M^{-1} \geqslant 0$ and $N \geqslant 0$, it follows that

$$M^{-1}N \geqslant 0$$

so that corresponding to the eigenvalue $\rho(M^{-1}N)$ the eigenvector is

$$\mathbf{x} \geqslant \mathbf{0}$$

which is also an eigenvector of H.

But $A^{-1} \geqslant 0$ (Lemma 5.1), hence

$$H = A^{-1}N \geqslant 0$$

so that corresponding to the eigenvector $\mathbf{x} \geqslant \mathbf{0}$ of H, the eigenvalue

$$\lambda \geqslant 0.$$

Any eigenvalue μ of $M^{-1}N$ is related to λ by (5.8) and for $\lambda \geqslant 0$, this is a steadily increasing function.

So the largest eigenvalue of $M^{-1}N$, that is $\rho(M^{-1}N)$, is obtained when the eigenvalue of $A^{-1}N$ is $\rho(A^{-1}N)$, and by (5.8)

$$\rho(M^{-1}N) = \frac{\rho(A^{-1}N)}{1 + \rho(A^{-1}N)} < 1.$$

It is of some interest to be able to compare the convergence of the iterative process for different splittings of the matrix A. If $A^{-1} > 0$, this can be achieved by comparing the spectral radii of the iteration matrices.

Lemma 5.10

If $A \in \mathcal{M}$ is such that $A^{-1} > 0$ and

$$A = M_1 - N_1$$
$$A = M_2 - N_2$$

are two convergent regular splittings of A, then

$$\rho(M_2^{-1}N_2) > \rho(M_1^{-1}N_1)$$

if

$$N_2 \geqslant N_1 \geqslant 0.$$

Proof

Since $A^{-1} > 0$ and $N_2 \geqslant N_1 \geqslant 0$, it follows that

$$A^{-1}N_1 \geqslant A^{-1}N_1 \geqslant 0.$$

Assuming that $A^{-1}N_1$ is irreducible

$$\rho(A^{-1}N_2) > \rho(A^{-1}N_1) > 0 \quad \text{(by Theorem 4.4)}$$

Since (5.8) shows that the eigenvalue $\rho(M^{-1}N)$ is an increasing function of $\rho(A^{-1}N)$, it follows that

$$\rho(M_2^{-1}N_2) > \rho(M_1^{-1}N_1).$$

Next we consider that $A^{-1}N_1$ is reducible.

We can transform it into a normal form (see Section 4.5.2). Such a transformation rearranges the diagonal entries, but they remain along the diagonal.

Since every entry of A^{-1} is positive, the (j, j)th entry of $A^{-1}N_1$ involves every entry of the jth column of N_1 ($j = 1, 2, \ldots, n$).

Since $N_2 \geqslant N_1$, at least one entry of N_2, say in the kth column is greater than the corresponding entry in N_1.

Hence the corresponding kth diagonal entry in $A^{-1}N_2$ is greater than the kth diagonal entry in $A^{-1}N_1$.

So in this case also

$$\rho(A^{-1}N_2) > \rho(A^{-1}N_1) > 0$$

and the conclusion follows.

PROBLEMS

5.1 Given

$$A = \begin{bmatrix} 4 & -2 & -1 \\ -1 & 3 & -1 \\ -1 & -2 & 4 \end{bmatrix}$$

show that

$$A^{-1} > 0$$

(Verification of Lemmas 5.1 and 5.4.)

5.2 Find the eigenvalues of

$$A = \begin{bmatrix} 2 & 0 & -1 \\ -2 & 3 & -1 \\ 0 & -1 & 1 \end{bmatrix}$$

and so verify Lemma 5.7.

5.3 For

$$B=\begin{bmatrix} 4 & -2 & 0 \\ 0 & 3 & -1 \\ -1 & 0 & 4 \end{bmatrix}$$

and A given in Problem 5.1 above so that $B \geqslant A$, show that (Lemma 5.6) $A^{-1} \geqslant B^{-1}$ and $|B| \geqslant |A| > 0$.

5.4 Write the matrix B in Problem 5.3 as a product of triangular matrices L and U (Lemma 5.7) such that $L, U \in \mathcal{M}$.

5.5 The matrix in Problem 5.3, can be written as

$$\begin{bmatrix} 4 & -2 & 0 \\ 0 & 3 & -1 \\ -1 & 0 & 4 \end{bmatrix} = \begin{bmatrix} 4 & -1 & 0 \\ 0 & 3 & 0 \\ 0 & -1 & 4 \end{bmatrix} - \begin{bmatrix} 0 & 1 & 0 \\ 0 & 0 & 1 \\ 1 & 1 & 0 \end{bmatrix} = M_1 - N_1 .$$

Verify that $B = M - N$ is a regular splitting and evaluate

$$\rho(M_1^{-1}N_1)$$

(see Lemma 5.9).

5.6 Writing the matrix B is Problem 5.3 as

$$\begin{bmatrix} 4 & -2 & 0 \\ 0 & 3 & -1 \\ -1 & 0 & 4 \end{bmatrix} = \begin{bmatrix} 4 & 0 & 0 \\ 0 & 3 & 0 \\ 0 & -1 & 4 \end{bmatrix} - \begin{bmatrix} 0 & 2 & 0 \\ 0 & 0 & 1 \\ 1 & 1 & 0 \end{bmatrix} = M_2 - N_2$$

so that $N_2 \geqslant N_1$ (in Problem 5.5).

Evaluate $\rho(M_2^{-1}N_2)$ and verify that

$$\rho(M_2^{-1}N_2) > \rho(M_1^{-1}N_1)$$

(Lemma 5.10).

6

Finite Markov Chains and Stochastic Matrices

The object of this chapter is to introduce some aspects of finite Markov chains and to show the relevance of the theory of nonnegative matrices to these concepts.

We consider a system which can be in exactly one of m states $S_1, S_2, \ldots, S_m$. It is assumed that the probability of transition from a state S_i to a state S_j in one unit of time (or one *step*) depends only on the two states involved.

Given that the system is in state S_i at step $(k-1)$, we denote by p_{ij}, called the *transition probability*, the probability that it will be in state S_j at step k. By the above assumption p_{ij} is independent of k so that the process is called a *homogeneous* Markov chain.

The matrix

$$P = [p_{ij}] \quad \text{of order } (m \times m)$$

is called the *transition matrix* for the *Markov chain*.

From the above assumptions it is clear that

$$\left.\begin{aligned} & p_{ij} \geqslant 0 \qquad i, j = 1, 2, \ldots, m \\ \text{and} \quad & \\ & \sum_{j=1}^{m} p_{ij} = 1 \qquad i = 1, 2, \ldots, m \end{aligned}\right\} \tag{6.1}$$

Definition 6.1
A matrix P satisfying conditions (6.1) is called (row) *stochastic*. If the matrix also satisfies

$$\sum_{i=1}^{m} p_{ij} = 1 \qquad j = 1, 2, \ldots, m$$

P is called *doubly stochastic.*

Theorem 6.1
(1) The spectral radius of a stochastic matrix P is 1.
(2) A nonnegative matrix P is stochastic if and only if the eigenvector corresponding to the eigenvalue 1 is

$$\mathbf{e} = [1, 1, \ldots, 1]'.$$

Proof
(1) Let λ be any eigenvalue of P, then by Gershgorin's theorem (Section 1.7) and (6.1)

$$|\lambda| \leqslant 1.$$

But

$$P\mathbf{e} = 1 \,.\, \mathbf{e}$$

hence,

$$\rho(P) = 1.$$

(2) If P is stochastic, Theorem 6.1 can be written as

$$P\mathbf{e} = 1 \,.\, \mathbf{e}$$

Hence 1 is an eigenvalue and $\mathbf{e}$ the corresponding eigenvector of P.
Conversely, if 1 is an eigenvalue of P and $\mathbf{e}$ the corresponding eigenvector, then

$$P\mathbf{e} = 1 \,.\, \mathbf{e}$$

which in turn implies (6.1).

Theorem 6.2
If P is a stochastic matrix then so is

$$P^k (k = 1, 2, \ldots).$$

Proof
Since P is stochastic, then

$$P\mathbf{e} = \mathbf{e}$$

so that

$$P^2\mathbf{e} = P(P\mathbf{e}) = P\mathbf{e} = \mathbf{e}$$

and by induction

$$P^k\mathbf{e} = \mathbf{e}\ (k = 1, 2, \ldots)$$

The result now follows by Theorem 6.1. ■

Writing

$$P^{(n)} = [p_{ij}^{(n)}]$$

we have

$$p_{ij}^{(2)} = \sum_k p_{ik}\, p_{kj}$$

which is the sum of the probabilities of transition along all possible paths from S_i to S_j of length 2. More generally

$$p_{ij}^{(n)} = \sum_k p_{ik}\, p_{kj}^{(n-1)} \tag{6.2}$$

which is the corresponding probability of transition along the paths of lengths n. If $p_{ij}^{(k)} > 0$ for some integer k, we will say that the S_j can be *reached* from the state S_i.

More precisely it will be said that S_j can be reached from S_i in k steps.

Definition 6.2

(1) A set of states C is *closed* (or *final*) if not state outside C can be reached from any state within C.

(2) S_j is called an *absorbing state* if it is the only state contained in a closed set C.

(3) The set C of states is called *irreducible* if C is closed and every state in C can be reached from every other state in C.

(4) A Markov chain defined on an irreducible set C is called *irreducible.*

The above definitions will allows us to decompose all the states associated with a Markov chain into two sets; *persistent* and *transient*.

To illustrate some of the definitions, consider the following example

$$P = \begin{array}{c} \begin{matrix} 1 & 2 & 3 & 4 & 5 & 6 & 7 & 8 \end{matrix} \\ \left[\begin{matrix} 1/3 & 0 & 0 & 0 & 2/3 & 0 & 0 & 0 \\ 0 & 1/3 & 0 & 1/3 & 1/3 & 0 & 0 & 0 \\ 0 & 0 & 1/2 & 0 & 0 & 0 & 0 & 1/2 \\ 0 & 0 & 0 & 1 & 0 & 0 & 0 & 0 \\ 1/2 & 0 & 0 & 0 & 1/2 & 0 & 0 & 0 \\ 1/8 & 0 & 1/4 & 1/8 & 0 & 1/4 & 1/4 & 0 \\ 0 & 0 & 0 & 0 & 0 & 0 & 1 & 0 \\ 0 & 0 & 1/4 & 0 & 0 & 0 & 0 & 3/4 \end{matrix}\right] \begin{matrix} 1 \\ 2 \\ 3 \\ 4 \\ 5 \\ 6 \\ 7 \\ 8 \end{matrix} \end{array}$$

The class of states that can be reached from a state S_j form a closed set. For example, take $j = 2$, the states that can be reached from S_2 form the set

$$C_1 = \{S_1, S_2, S_4, S_5\}$$

which is closed.

But S_4 is an absorbing state, and once in S_4 the system cannot enter any other state. It follows that C_1 is not irreducible. On the other hand for $j = 3$, we obtain the set

$$C_2 = \{S_3, S_8\}$$

which is irreducible.

It is clear that for a set C to be irreducible it must be closed and must contain no proper closed subsets.

We can transform P by an appropriate permutation of the rows and columns, for example by reordering the index set so that the indices corresponding to the absorbing states come first followed by the indices corresponding to the irreducible sets. On taking the index set in the order $\{4, 7, 1, 5, 3, 8, 2, 6\}$, the above matrix is transformed into

$$\begin{array}{cc:cc:cc|cc|c}
4 & 7 & 1 & 5 & 3 & 8 & 2 & 6 & \\
1 & 0 & 0 & 0 & 0 & 0 & 0 & 0 & 4 \\
0 & 1 & 0 & 0 & 0 & 0 & 0 & 0 & 7 \\
\hdashline
0 & 0 & 1/3 & 2/3 & 0 & 0 & 0 & 0 & 1 \\
0 & 0 & 1/2 & 1/2 & 0 & 0 & 0 & 0 & 5 \\
\hdashline
0 & 0 & 0 & 0 & 1/2 & 1/2 & 0 & 0 & 3 \\
0 & 0 & 0 & 0 & 1/4 & 3/4 & 0 & 0 & 8 \\
\hdashline
1/3 & 0 & 0 & 1/3 & 0 & 0 & 1/3 & 0 & 2 \\
1/4 & 1/4 & 1/8 & 0 & 0 & 1/4 & 0 & 1/8 & 6
\end{array}$$

The transformed matrix has the standard reducible form

$$\begin{bmatrix} P_{11} & 0 \\ P_{21} & P_{22} \end{bmatrix}$$

where P_{11} is associated in the above example with (at least) three closed sets

$$\{S_4, S_7\}, \quad \{S_1, S_5\}, \quad \text{and} \quad \{S_3, S_8\}$$

each a proper subset of the closed set $\{S_1, S_3, S_4, S_5, S_7, S_8\}$. By the definition of a closed set C, the probability that the system assumed in state $S_i \in C$ will be at a subsequent time in state $S_j \in C'$ (where C' is the complement of C) is zero.

Hence

$$p_{ij}^{(n)} = 0 \qquad n = 1, 2, \ldots \tag{6.3}$$

whenever $S_i \in C$ and $S_j \in C'$.

This result can be used to transform a transition matrix into the above standard form.

Another (related) important consequence of closure is that the submatrix associated with a closed set C is stochastic.

For example, the submatrix P_{11} in the above example, associated with the closed set

$$C = \{S_4, S_7, S_1, S_5, S_3, S_8\}$$

is stochastic.

This is also true for the submatrix associated with each proper closed subset of C.

This result leads to the following important criterion.

On deleting the rows and columns corresponding to the states *not* contained in a closed set C from the matrix P^n (see Theorem 6.2), the submatrix remaining is stochastic.

Conversely, it must be true that on deleting the rows and columns from P^n corresponding to the states not contained in a set C which is *not* closed, the submatrix remaining is *substochastic* which is a matrix whose row sums $\leqslant 1$. ■

We now define a classification of the states which is useful when considering the behaviour of P^n as $n \to \infty$.

Let $f_{ij}^{(n)}$ be the probability that the system initially at S_i first enters S_j at the nth step. We define

$$f_{ij}^{(0)} = 0.$$

Let $p_{jj}^{(n)}$ be the transition probability from S_j to S_j in n steps.

The probability that the systems first returns to S_j in k steps and then again at step n, where $n > k$ is

$$f_{jj}^{(k)}\, p_{jj}^{(n-k)}.$$

Since we are dealing with mutually exclusive events, it follows that

$$p_{jj}^{(n)} = \sum_{k=1}^{n} f_{jj}^{(k)}\, p_{jj}^{(n-k)} \qquad n \geqslant 1, \tag{6.4}$$

where we define

$$p_{jj}^{(0)} = 1.$$

The probability that starting from S_j, the system will ever return to S_j is

$$f_{jj} = \sum_{n=1}^{\infty} f_{jj}^{(n)}. \tag{6.5}$$

If a return is certain, that is if $f_{jj} = 1$, then

$$\mu_j = \sum_{n=1}^{\infty} n f_{jj}^{(n)} \tag{6.6}$$

is the *mean recurrence* time of S_j.

Definition 6.3

The state S_j is called *persistent* (or *recurrent*) if $f_{jj} = 1$ and is *transient* if $f_{jj} < 1$. Notice that if a state is transient, there is a positive probability that the system will never return to it.

Definition 6.4

S_j is *periodic* with *period d* if the system cannot return to S_j except perhaps at step $d, 2d, 3d, \ldots,$ and d is the largest integer (> 1) having this property.

If S_j is not periodic (that is if $d = 1$), S_j is called *aperiodic*.

An aperiodic persistent state S_j whose mean recurrence time $\mu_j = \infty$ is called a *null state*.

Theorem 6.3

Using the notation defined above:

(1) If

$$\sum_{n=0}^{\infty} p_{jj}^{(n)}$$

is finite, then S_j is transient.

(2) If

$$\sum_{n=0}^{\infty} p_{jj}^{(n)} = \infty,$$

then S_j is persistent.

Proof

We note that in (6.4), the sequence $\{p_{jj}^{(n)}\}$ is a convolution of the sequence $\{f_{jj}^{(n)}\}$ and $\{p_{jj}^{(n)}\}$. Define the generating functions

$$F(s) = \sum_{n=1}^{\infty} f_{jj}^{(n)} s^n,$$

and

$$G(s) = \sum_{n=0}^{\infty} p_{jj}^{(n)} s^n.$$

The generating function of the left-hand side of (6.4) is

$$G(s) - 1$$

since the term $p_{jj}^{(0)}$ (= 1) is missing.

We obtain the generating function of the right-hand side of (6.4), by using the Convolution Theorem, so that

$$G(s) - 1 = F(s)\,G(s)$$

or

$$G(s) = \frac{1}{1 - F(s)}. \tag{6.7}$$

Now

$$\lim_{s \to 1} F(s) = \sum_{n=1}^{\infty} f_{jj}^{(n)} = f_{jj} \qquad \text{(see (6.5))}.$$

As s increases to 1, $G(s)$ increases monotonically, hence

$$\lim_{s \to 1} G(s) = \lim_{s \to 1} \sum_{n=0}^{\infty} p_{jj}^{(n)} s^n \leqslant \sum_{n=0}^{\infty} p_{jj}^{(n)}. \tag{6.8}$$

From (6.6) and the above, we conclude that

$$\lim_{s \to 1} G(s) = \left.\begin{matrix} (1 - f_j)^{-1} \\ \infty \end{matrix}\right\} \begin{matrix} \text{if } f_{jj} < 1 \\ \text{if } f_{jj} = 1 \end{matrix}$$

The theorem now follows from Definition 6.3 and (6.8). ∎

We now return to consider an irreducible Markov chain defined on a set C. Let $S_j, S_k \in C$.

N be the shortest path from S_j to S_k, and

M be the shortest path from S_k to S_j.

Let

$$p_{jk}^{(N)} = \alpha > 0 \quad \text{and} \quad p_{kj}^{(M)} = \beta > 0$$

then for any $n > 0$

$$\left.\begin{aligned} p_{jj}^{(n+N+M)} &\geqslant p_{jk}^{(N)}\, p_{kk}^{(n)}\, p_{kj}^{(M)} = \alpha\beta p_{kk}^{(n)} \\ p_{kk}^{(n+N+M)} &\geqslant p_{kj}^{(M)}\, p_{jj}^{(n)}\, p_{jk}^{(N)} = \alpha\beta p_{jj}^{(n)} \end{aligned}\right\} \tag{6.9}$$

which shows that the asymptotic behaviour of $p_{kk}^{(n)}$ and $p_{jj}^{(n)}$ is the same. This is another way of saying that the states forming an irreducible chain C are all of the same type. For example if S_j is persistent, so that

$$\sum_{n=0}^{\infty} p_{jj}^{(n)} = \infty$$

(6.9) shows that both

$$\sum p_{jj}^{(n)} \quad \text{and} \quad \sum p_{kk}^{(n)}$$

are divergent. It follows that all states of a (finite) irreducible Markov chain are either all persistent or (by a similar argument) all transient.

Also assume that, S_j is periodic with period d.

Since for $n = 0$, the right-hand side of (6.9) is positive $(N + M)$ is divisible by d.

Thus (6.9) shows that S_k is periodic with period d. Again, it follows that if one state is periodic, then all states in an irreducible chain are periodic.

It can be shown that if S_j is persistent and periodic, then

$$\lim_{n\to\infty} p_{jj}^{(n)} = \frac{d}{\mu_j} \quad \text{if } n \text{ is divisible by } d$$

$$= 0 \quad \text{otherwise.}$$

Where μ_j is the mean recurrence time (see (6.6)) and d is the period.

We shall see (Theorem 6.5) that a finite Markov chain does not involve null states (periodic or otherwise). Hence in this case

$$p_{jj}^{(n)} > 0 \quad \text{if } n \text{ is divisible by } d,\ n > N \tag{6.10}$$

Let C be an irreducible set of periodic states. All states in C have the same period d (say).

Since C is irreducible, every state in C can be reached from every other state.

Let $S_j \in C$.

$$\exists u,\ v \in Z^+$$

such that

$$p_{1j}^{(u)} > 0$$

and

$$p_{j1}^{(v)} > 0,$$

so that

$$p_{jj}^{(u+v)} \geqslant p_{j1}^{(v)}\, p_{1j}^{(u)} > 0.$$

It follows by (6.1) that $(u + v)$ is divisible by d.

Keeping v fixed, it is seen that for

$$p_{1j}^{(u)} > 0$$

u must be of the form

$$u = qd + r_j$$

where

$$0 \leqslant r_j < d$$

is a characteristic of the state S_j.

Since S_j is any state in C, it follows that all the states can be partitioned into d mutually exclusive sets $C_0, C_1, \ldots, C_{d-1}$ so that if $m = qd + r_j$

$$p_{ij}^{(m)} > 0 \quad \text{if } S_j \in C_{r_j}$$

$$= 0 \text{ otherwise}$$

More generally

$$p_{ij}^{(md)} > 0 \text{ if } S_j \in C_{r_j}$$

$$= 0 \text{ otherwise.}$$

For a persistent null state S_j, $\mu_j = \infty$, so that

$$p_{jj}^{(n)} \to 0 \quad \text{as} \quad n \to \infty.$$

Definition 6.5

An aperiodic persistent non-null ($u_j < \infty$) state S_j is called *ergodic*.

It can be shown that for an ergodic state S_j

$$p_{jj}^{(n)} \to \alpha_j, p_{ij}^{(n)} \to \beta_j \quad \text{as} \quad n \to \infty. \tag{6.11}$$

where

$$\alpha_j = 1/\mu_j \quad \text{and} \quad \beta_j = \frac{f_{ij}}{\mu_j}.$$

On the other hand, using Theorem 6.3, it is not difficult to prove that if S_j is a persistent null state, then

$$p_{jj}^{(n)} \to 0 \quad \text{and} \quad p_{ij}^{(n)} \to 0 \text{ is } n \to \infty \tag{6.12}$$

Theorem 6.4

Let $C = \{S_i\}$ where S_i is a state that can be reached from a persisten state S_j. Then C is irreducible.

Proof

By Definition 6.3, S_j can be reached from S_i. If S_i, $S_k \in C$, then (by the above argument) S_j can be reached from S_k, hence S_i can be reached from S_k (Fig. 8).

By a similar argument, every state in C can be reached from every other state in C. The lemma follows (see Definition 6.2).

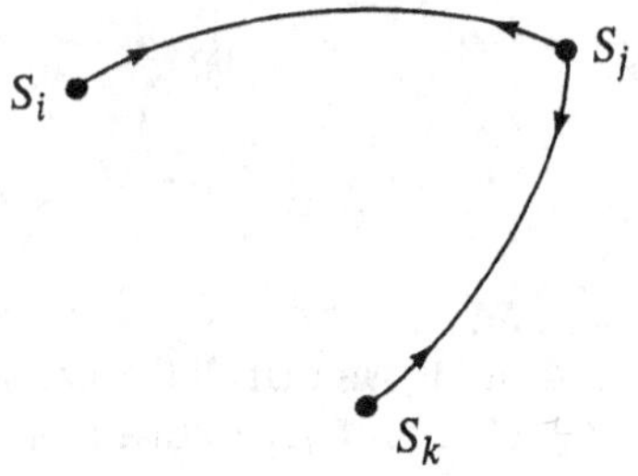

Fig. 8

Theorem 6.5
In a finite Markov chain

(1) not all states can be transient, and
(2) there can be no persistent null states.

Proof
(1) Let

$$P^n = [p_{ij}^{(n)}].$$

If any diagonal term, say $p_{kk}^{(n)}$ of P^n is positive and not decreasing to 0 as $n \to \infty$, then

$$\sum_{n=0}^{\infty} p_{kk}^{(n)} = \infty$$

as S_k is a persistent state (see Theorem 6.3) proving Part (1) of the theorm.

So assume all diagonal entries of P^n are 0. By Theorem 6.3, P^n is stochastic, so that it is not possible for

$$p_{ij}^{(n)} \to 0 \quad \text{for all } (i, j).$$

In fact it is claimed that for some (i, j) and $m > N$

$$p_{ij}^{(m)} = \alpha > 0 \quad \text{and} \quad p_{ji}^{(m)} = \beta > 0.$$

For assume on the contrary, that

$$p_{ji}^{(m)} = 0 \ \text{ all } (i, j).$$

Then P^m is strictly upper triangular, with all zero entries in its last row, which is impossible since P^m is stochastic.

So $p_{ji}^{(m)} > 0$, and it follows that

$$p_{jj}^{(2m)} \geqslant p_{ji}^{(m)} p_{ij}^{(m)}$$

showing that S_j is a persistent state.

Part 1 of the theorem follows.

(2) Assume that S_j is a persistent null state. Then by Theorem 6.4, $C = \{S_i\}$ where the states S_i can be reached from S_j, is irreducible.

On deleting the rows of columns of P corresponding to the states not included in C, there remains a stochastic matrix Q (say), corresponding to the null states only.

Then

$$Q = [q_{ij}]$$

such that

$$q_{ij}^{(n)} \to 0$$

as $n \to \infty$ (all i, j).

Which implies that $Q^n \to 0$ or that Q is not stochastic.

Hence the assumption that S_j is a persistent null state is false. Part (2) of the theorem follows.

Theorem 6.4 applies only to persistent states if a *finite* Markov chain is considered. If S_j is a transient state, the set $C = \{S_i\}$, where S_i can be reached from S_j, is not irreducible. Indeed, it was already shown in the above illustrative example that the set of states reachable from S_2, $C = \{S_1, S_2, S_4, S_5\}$ is not irreducible. This implies that no transient state can be reached from a persistent one so that if a finite Markov chain contains both transient and persistent states, the corresponding transition matrix P can be partitioned into the form (see the illustrative example):

$$\begin{array}{c} \begin{array}{ccc} \overbrace{}^{C_1} & \overbrace{}^{C_2} & \overbrace{}^{C_3} \end{array} \\ \left[\begin{array}{cc:c} I & 0 & 0 \\ 0 & P_1 & 0 \\ \hdashline P_2 & P_3 & P_{22} \end{array}\right] \begin{array}{l} \} \, C_1 \\ \} \, C_2 \\ \} \, C_3 \end{array} \end{array}$$

where

- C_1 is the set of absorbing states.
- C_2 is a set of persistent states which can be uniquely partitioned into irreducible sets (if there is more than one such a set).
- C_3 is a set of transient states.
- P_1 is a stochastic matrix of transition probabilities within the persistent non-absorbing states.
- P_2 is a matrix of transition probabilities from the transient to the absorbing states.
- P_3 is a matrix of transition probabilities from the transient to the persistent non-absorbing states.
- P_{22} is a matrix of transition probabilities within the transient states.

The elegance of the theory of finite Markov chains depends partly on the fact that the nature of the chain can be determined when the eigenvalues of the associated transition matrix P are known.

By considering various types of matrices associated with Markov chains defined on a single class of states, the type of system can be determined. Obviously the converse can also be attempted so that given a Markov chain the type of the associated transition matrix can be determined.

Before carrying out this exercise we consider, as an example, the following interesting result.

Example 6.1

(1) Consider a nonnegative and irreducible matrix $P(n \times n)$ such that $\rho(P) = 1$, and corresponding to the eigenvalue 1, $\mathbf{x}$ and $\mathbf{y}$ are the eigenvectors of P and P' respectively. Show that

$$\lim_{k\to\infty} P^k = (\mathbf{y}'\mathbf{x})^{-1}\mathbf{x}\mathbf{y}'$$

(2) If P is stochastic matrix, show that

$$\lim_{k\to\infty} P^k = \mathbf{e}\,.\,\mathbf{z}'$$

where $\mathbf{z}'$ is a stochastic vector (sum of components is 1).

Solution

From Section 1.14 and Chapter 4, we know that P is power convergent.

Let

$$\lim_{k\to\infty} P^k = Q = [\mathbf{q}_1, \mathbf{q}_2, \ldots, \mathbf{q}_n] \quad \text{(say)}.$$

Also

$$Q = \lim_{k\to\infty} P^{k+1} = P \lim_{k\to\infty} P^k = PQ,$$

so that

$$\mathbf{q}_j = P\mathbf{q}_j \qquad (j = 1, 2, \ldots, n).$$

But

$$P\mathbf{x} = 1\,.\,\mathbf{x}$$

so that

$$\mathbf{q}_j = r_j\mathbf{x}$$

where r_j is a scalar.

Hence

$$\begin{aligned} Q &= [r_1 x_1, r_2 x_2, \ldots, r_n x_n] \\ &= \mathbf{x}\,\mathbf{r}' \end{aligned}$$

where

$$\mathbf{r} = [r_1, r_2, \ldots, r_n]', \quad \text{and} \quad \mathbf{x} = [x_1, x_2, \ldots, x_n]' \quad p_{ij}^{(n)}$$

Also

$$P'\mathbf{y} = \mathbf{y}$$

so that

$$(P')^k \mathbf{y} = \mathbf{y} \qquad (k = 1, 2, \ldots),$$

it follows that

$$\mathbf{y}'Q = \mathbf{y}' \lim_{k\to\infty} P^k = \lim_{k\to\infty} \mathbf{y}'P^k = \mathbf{y}'$$

or

$$\mathbf{y}'\mathbf{x}\,\mathbf{r}' = \mathbf{y}'$$

so

$$\mathbf{y}'\mathbf{x} \neq 0.$$

Hence

$$\mathbf{r}' = (\mathbf{y}'\mathbf{x})^{-1}\mathbf{y}'$$

and

$$Q = (\mathbf{y}'\mathbf{x})^{-1}\mathbf{x}\,\mathbf{y}',$$

the result follows.

If P is stochastic, then we can choose

$$\mathbf{x} = \mathbf{e}$$

so that

$$Q = (\mathbf{y}'\mathbf{e})^{-1}\mathbf{e}\,\mathbf{y}' = \mathbf{e}(\mathbf{y}'\mathbf{e})^{-1}\mathbf{y}' = \mathbf{e}\,\mathbf{z}'$$

where

$$\mathbf{z}' = (\mathbf{y}'\mathbf{e})^{-1}\mathbf{y}'$$

is a stochastic vector.

Notice that if in the above calculations $\mathbf{x}$ and $\mathbf{y}$ are chosen to be normalised in such a way that

$$\mathbf{y}'\mathbf{x} = 1$$

then the above result can be written in the simplified form as

$$Q = \lim_{k\to\infty} P^k = \mathbf{x}\,\mathbf{y}' \tag{6.13}$$

Example 6.2
Given

$$P = \begin{bmatrix} 0.8 & 0.1 & 0.1 \\ 0.1 & 0.8 & 0.1 \\ 0.1 & 0.1 & 0.8 \end{bmatrix}$$

find

$$\lim_{k\to\infty} P^k.$$

Solution
P (and P') is a stochastic matrix, hence

$$\mathbf{x} = \mathbf{y} = \mathbf{e}.$$

and

$$\mathbf{z}' = \frac{1}{3}\,[1 \quad 1 \quad 1],$$

so that

$$Q = \begin{bmatrix} 1/3 & 1/3 & 1/3 \\ 1/3 & 1/3 & 1/3 \\ 1/3 & 1/3 & 1/3 \end{bmatrix}. \qquad \blacksquare$$

We are now ready to consider various types of transition matrices.

P is primitive and irreducible.
Since P also is stochastic

$$\lim_{k\to\infty} P^k = \mathbf{e}\,\mathbf{z}'$$

where

$$\mathbf{z}' = [z_1, z_2, \ldots, z_m] > \mathbf{0}$$

Hence

$$p_{ij}^{(n)} \to z_j > 0 \text{ is } n \to \infty$$

from which it follows that the corresponding Markov chain is ergodic (see 6.11)).

Conversely if the chain is ergodic, then

$$p_{ij}^{(n)} > 0$$

all (i, j) for $n > k$ (see (6.11)).

It follows that

(1) $P = [p_{ij}]$ is primitive.
(2) $G(P)$, the digraph of P is strongly connected, so that P is irreducible.

P is irreducible but primitive.
In this case P is irreducible and cyclic with $d > 1$ (say) eigenvalues of unit modulus.

It then follows from the discussion in Section 4.5 and above that the corresponding chain contains only periodic states each with period d.

P is reducible.
We consider $P(n \times n)$ in the standard form

$$P = \begin{bmatrix} \overbrace{P_{11}}^{C} & \overbrace{0}^{T} \\ P_{21} & P_{22} \end{bmatrix} \begin{matrix} \} C \\ \} T \end{matrix}$$

where P_{11}, P_{21} and P_{22} are non-zero.

If P_{11} is of order $m \times m$, there are m states contained in the associated closed set C. Then P_{22} is of order $(n-m) \times (n-m)$ and there are $(n-m)$ states contained in the set T associated with P_{22}.

Initially we shall assume that P_{11} is irreducible (which is not the case for the above illustrative example). Further, we assume that P_{11} is primitive.

These assumptions imply that one eigenvalue P_{11} is 1 and the remaining $(m-1)$ eigenvalues are in modulus smaller than 1.

Since $P_{21} \neq 0, P_{22}$ is a substochastic matrix, so the moduli of all the eigenvalues of P_{22} are smaller than 1, from which it follows that

$$P_{22}^n \to 0 \text{ as } n \to \infty.$$

It follows that there exists a non-singular matrix U such that

$$UPU^{-1} = \begin{bmatrix} 1 & 0 \\ 0 & J \end{bmatrix}$$

where J is a Jordan matrix of order $(n-1) \times (n-1)$ having all eigenvalues smaller than 1 in modulus. Since

$$P^n = U^{-1} \begin{bmatrix} 1 & 0 \\ 0 & J^n \end{bmatrix} U^{-1}$$

and $J^n \to 0$ as $n \to \infty$ (see Section 1.14), it follows that

$$\lim_{n \to \infty} P^n$$

exists.

Using the standard partitioned form of P, we have

$$P^n = \begin{bmatrix} P_{11}^n & 0_n \\ Q_n & P_{22} \end{bmatrix}$$

where Q_n is a function of P_{11}, P_{12} and P_{22}. As we have seen (Example 6.1)

$$\lim_{n\to\infty} P_{11}^n = \mathbf{e}\mathbf{z}'$$

where $\mathbf{e}$ and $\mathbf{z}$ are each of order m.

Since P is convergent

$$Q_n \to Q \text{ (say) as } n \to \infty.$$

Hence

$$\lim_{n\to\infty} P^n = \begin{bmatrix} \mathbf{e}\mathbf{z}' & 0 \\ Q & 0 \end{bmatrix}.$$

We shall now evaluate Q.

Since the row sums of the matrix $[P_{21} + P_{22}]$ are 1, we have

$$[P_{21} + P_{22}]\mathbf{e} = \mathbf{e}.$$

P_{22} has no eigenvalue equal to 1, so $[I - P_{22}]$ is non-singular, so that

$$[I - P_{22}]^{-1} P_{21}\, \mathbf{e}(m) = \mathbf{e}(n - m)$$

where $\mathbf{e}(j)$ indicates the one vector of order j.

Writing

$$P^{n+1} = PP^n$$

we obtain

$$Q_{n+1} = P_{21}P_{11}^n + P_{22}Q_n.$$

As $n \to \infty$

$$Q = P_{21}\, \mathbf{e}\mathbf{z}' + P_{22}Q$$

or

$$Q = [I - P_{22}]^{-1} P_{21}\, \mathbf{e}\mathbf{z}' = \mathbf{e}(n - m)\, \mathbf{z}'(m).$$

It follows that

$$\lim_{n\to\infty} P^n = \begin{bmatrix} \mathbf{e}\mathbf{z}' & 0 \\ \mathbf{e}\mathbf{z}' & 0 \end{bmatrix} \begin{matrix} \} m \\ \} n-m \end{matrix} = \begin{bmatrix} 1 \\ 1 \\ \vdots \\ 1 \end{bmatrix} [z_1 \ldots z_m\, 0 \ldots 0].$$

So if $S_j \in C$, then

$$\lim_{n\to\infty} p_{jj}^{(n)} = z_j > 0$$

so that S_j is ergodic (see Definition 6.5).

On the other hand if $S_j \in T$, then as $n \to \infty$ $p_{jj}^{(n)}$ decreases to 0, the rate of convergence is of order $O(r^n)$ where $0 < r < 1$, r being the largest eigenvalue of P_{22}.

It follows that

$$\sum_{n=1}^{\infty} p_{jj}^{(n)} < \infty$$

so that S_j is transient.

If P_{11} is itself in a block diagonal form

$$P_{11} = \begin{bmatrix} P_1 & \dots & 0 \\ \vdots & & \vdots \\ 0 & \dots & P_r \end{bmatrix}$$

where P_j $(j = 1, 2, \dots, r)$ are each irreducible, then we have a situation similar to the one in the illustrative example.

It corresponds to the closed set C associated with P_{11} containing a number of proper closed subsets.

Again it is assumed that the stochastic matrices $P_1, P_2, \dots, P_r$ are primitive in which case

$$\lim_{n\to\infty} P_{11}^n = \begin{bmatrix} Q_1 & \dots & 0 \\ \vdots & & \vdots \\ 0 & \dots & Q_r \end{bmatrix} = Q \text{ (say)}$$

where

$$Q_j = \lim_{n\to\infty} P_j^n > 0.$$

Using an argument analogous to the above, we conclude that

$$C = \bigcup_{s=1}^{r} C_s$$

where

$$C_i \cap C_j = \varnothing$$

where C_s $(s = 1, \dots, r)$ is a closed set of ergodic states.

If P_{11} (or one of the submatrices P_s) is not primitive, then it has more than one eigenvalue of unit modulus.

It follows (see Section 1.14) that the matrix P is not convergent, in fact in this case P_{11} is cyclic.

When one of the submatrices of P_s is cyclic and P is not convergent we can no longer use the above type of analysis which depends on P being convergent. But there are methods by which the periodic phenomena can be 'smoothed out', and results in a matrix tending to a limit. For further information see [B3].

7

Some Applications of Nonnegative Matrices

7.1 INTRODUCTION

In this chapter we consider and analyse various Problems or Models which can be analysed with the help of nonnegative matrices. This is an opportunity to illustrate some of the properties of such matrices developed over various chapters in this book. All the Models considered are of processes on a relatively small scale allowing the calculations of solutions without the use of a computer. Similar problems on a much larger scale can be solved using identical techniques.

The problems discussed in this chapter have been divided into three sections; Genetic, Economic and Markov Chain models. This division is rather arbitrary since, for example, the Genetic models could have been discussed in the Markov Chain section.

7.2 SOME GENETIC MODELS

Mendel's plant breeding experiments have resulted in the formulation of the Mendelian laws which are essentially probabilistic in nature. We are in essence interested in the propagation of various characteristics in plants and animals over a number of generations. If we can express this propagation in terms of matrices, these matrices must be nonnegative since we are dealing with a probabilistic phenomenon.

To understand these models and to define the terminology used we shall discuss briefly certain aspects of Mendel's Theory.

Plants and animals are composed of cells. Each cell contains a nucleus which contains structures called *chromosomes.* Chromosomes carry the hereditary factors or *genes*. Each individual possesses two genes which determine a specific characteristic. At mating each parent transmits only one of the genes to the offspring with probability 1/2.

Alleles (short for allemorphs) are genes responsible for contrasting characters. For example, in man there are two alleles for eye colouration, one for brown colour and one for blue colour.

There are two types of chromosomes:

(1) sex chromosomes and
(2) autosomes.

Humans have 46 chromosomes arranged in pairs in each nucleus; 2 sex chromosomes and 44 autosomes.

The female sex chromosomes are denoted by XX.

The male sex chromosomes are denoted by XY.

In a female each ovum carries an X chromosome. In a male each sperm carries either an X or a Y chromosome with equal probability.

As already pointed out, an individual possess two alleles associated with each physical characteristic. Frequently the two alleles are designated by the letters A and a.

If the two alleles are the same (AA or aa) the individual is said to *homozygous*, and if they are different (Aa) the individual is said to be *heterozygous.*

The term *genotype* denotes the individual's genetic constitution. *Phenotype* refers to the appearance of an individual which depends on his genotype and the effects of the environment.

Gamete formation is the result of sperm and ovum formation during which the number of chromosomes is halved so that each gamete has one member of each chromosome pair. At fertilisation, two gametes unite. Since each gamete contains a set of 23 chromosomes, the resulting fertilised cell (*zygote*) has 46 chromosomes.

There are certain characteristics or traits which are said to be *dominant* whereas others are said to be *recessive.*

For example, if one parent has brown eyes the other has blue eyes and their offsprings have brown eyes, then brown is said to be a dominant characteristic, whereas blue is a recessive one. Similarly if a tall plant is crossed with a dwarf one resulting in a first generation hybrid's being tall plants, then the tallness of plants is a dominant trait whereas the dwarfness is a recessive one.

We are now ready to discuss setting up a genetic model.

Consider humans with alleles A and a where for example A is 'brown-eyed' and a is 'blue-eyed'. We have three genotypes:

AA, Aa, aa

where no distinction is made between Aa and aA.

Since brown colour (A) is dominant, there is no difference in appearance between individuals having genotypes AA and Aa, both have brown eyes. The individual with genotype aa has blue eyes.

It has been explained above that if one parent is of genotype Aa, an offspring is equally likely to inherit the A allele or the a allele from that parent. On the other hand, if the parent is of genotype AA (or aa), an offspring will always inherit the A (or a) allele from that parent.

We can therefore draw up a table showing the probabilities of the genotypes of an offspring for all possible combinations of genotypes of the parents.

For example, if the genotypes of parents are

Aa and aa

then the probabilities of the possible genotypes of an offspring are

$$P(aa) = \frac{1}{2}, \quad P(Aa) = \frac{1}{2} \quad \text{and} \quad P(AA) = 0.$$

The full list of probabilities is shown in Table 1.

Table 1

	Genotypes of parents					
	aa–aa	aa–Aa	aa–AA	Aa–Aa	AA–Aa	AA–AA
Probabilities associated with an offspring						
aa	1	$\frac{1}{2}$	0	$\frac{1}{4}$	0	0
Aa	0	$\frac{1}{2}$	1	$\frac{1}{2}$	$\frac{1}{2}$	0
AA	0	0	0	$\frac{1}{4}$	$\frac{1}{2}$	1

To make further advance towards setting up genetic models we must make assumptions about the proportions of the genotypes in the population concerned and about the mating habits of this population.

For example, we may consider that mating takes place either randomly or, in contrast, an individual of one genotype always mates with an individual of the same genotype, in other words, an inbreeding process.

To discuss a most general example consider the alleles A and a for which the three genotypes AA, Aa, aa give rise to three distinct phenotypes. In this case A is not necessarily a dominant allele so that for garden peas, A may be 'red blossom' a may be 'blue blossom' and Aa may be 'pink blossom' (not 'red blossom' which would be the case if A was dominant).

We assume that the proportions of the three phenotypes in the population resulting from the genotypes AA, Aa, aa are initially p_0, q_0, r_0 respectively.

Model 1

If random mating takes place, we can calculate the probability that an offspring is of type AA (say). This involves calculating the probability of mating between two specified genotypes and the conditional probability of an offspring of type AA given the parental types.

We need only consider parental types (for this example) having at least one allele A each.

Table 2

Genotype of parent 1	Genotype of parent 2	Probability of mating	Conditional probability of offspring of type AA	Probability of offspring A
Aa	Aa	q_0^2	$\frac{1}{4}$	$\frac{1}{4}q_0^2$
Aa	AA	p_0q_0	$\frac{1}{2}$	$\frac{1}{2}p_0q_0$
AA	Aa	p_0q_0	$\frac{1}{2}$	$\frac{1}{2}p_0q_0$
AA	AA	p_0^2	1	p_0^2

Since the matings are independent, the third column is the product of probabilities in the first two columns.

The conditional probabilities in the fourth column are the ones already found in the last row of Table 1. The probabilities in the final column are the products of the probabilities in columns three and four.

Finally, since the events defined in the four rows of Table 2 are independent the probability of an offspring of type AA is the sum of the probabilities in the fifth column.

$$\begin{aligned} P(AA) &= \frac{1}{4}q_0^2 + p_0q_0 + p_0^2 \\ &= \left(p_0 + \frac{1}{2}q_0\right)^2 = p_1 \quad \text{(say).} \end{aligned}$$

Similarly

$$P(Aa) = 2\left(p_0 + \frac{1}{2}q_0\right)\left(r_0 + \frac{1}{2}q_0\right) = q_1 \quad \text{(say)}$$

and

$$P(aa) = \left(r_0 + \frac{1}{2}q_0\right)^2 = r_1 \quad \text{(say)}$$

Notice that the proportion of the A-alleles in the original population for which

$$p_0 + q_0 + r_0 = 1$$

is

$$p_0 + \frac{1}{2}q_0.$$

Also the proportion of the A-alleles in the first generation for which

$$p_1 + q_1 + r_1 = 1$$

is

$$p_1 + \frac{1}{2}q_1.$$

Now let

$$\lambda = p_0 + \frac{1}{2}q_0,$$

then

$$p_1 = \lambda^2, \quad q_1 = 2\lambda(1-\lambda) \quad \text{and} \quad r_1 = (1-\lambda)^2.$$

So the proportion of the A-alleles in the first generation is

$$p_1 + \frac{1}{2}q_1 = \lambda^2 + \lambda(1-\lambda) = \lambda,$$

which is the proportion of A-alleles in the original population.

In general, in this case

$$p_{n+1} = p_n, \; q_{n+1} = q_n \; r_{n+1} = r_n$$

or

$$\mathbf{x}^{(n+1)} = A\,\mathbf{x}^{(n)} \qquad n = 0, 1, 2, \ldots$$

where

$$\mathbf{x}^{(n)} = [p_n, q_n, r_n]' \quad (n = 0, 1, 2, \ldots)$$

and

$$A = \begin{bmatrix} 1 & 0 & 0 \\ 0 & 1 & 0 \\ 0 & 0 & 1 \end{bmatrix}.$$

More generally,

$$\mathbf{x}^{(n)} = A^n \mathbf{x}_0$$

where in this case A is the unit matrix (hence a nonnegative matrix).

It is seen that for this model the proportions of genotypes remain constant for all generations.

Model 2

We next consider a situation involving inbreeding in which mating takes place only between individuals of the same genotype.

Again let

p_n = proportion of population of genotype AA is the nth generation.

q_n = proportion of population of genotype Aa in the nth generation.

r_n = proportion of population of genotype aa in the nth population.

We have

$$p_n + q_n + r_n = 1 \qquad n = 0, 1, 2, \ldots .$$

Again looking at the last row of Table 1, we have

$$p_1 = p_0 + \frac{1}{4} q_0$$

and from the second row

$$q_1 = \frac{1}{2} q_0$$

and from the first row

$$r_1 = r_0 + \frac{1}{4} q_0 .$$

More generally

$$p_{n+1} = p_n + \frac{1}{4} q_n$$

$$q_{n+1} = \frac{1}{2} q_n$$

$$r_{n+1} = r_n + \frac{1}{4} q_n \qquad n = 0, 1, 2, \ldots$$

So that

$$\mathbf{x}^{(n+1)} = A\,\mathbf{x}^{(n)}$$

or

$$\mathbf{x}^{(n)} = A^n \mathbf{x}^{(0)}$$

where

$$\mathbf{x}^{(n)} = [p_n, q_n, r_n]'$$

and

$$A = \begin{bmatrix} 1 & \frac{1}{4} & 0 \\ 0 & \frac{1}{2} & 0 \\ 0 & \frac{1}{4} & 1 \end{bmatrix}.$$

Here we have a nonnegative (column stochastic) matrix A which is obviously reducible. Hence Theorem 4.9 applies.

In fact, the eigenvalues of the matrix are the eigenvalues of the matrices

$$[1] \quad \text{and} \quad \begin{bmatrix} \frac{1}{2} & 0 \\ \frac{1}{4} & 1 \end{bmatrix}$$

so that

$$\lambda_1 = 1,\ \lambda_2 = 1 \quad \text{and} \quad \lambda_3 = \frac{1}{2}.$$

The corresponding eigenvectors can be chosen as

$$\mathbf{x}_1 = \begin{bmatrix} 1 \\ 0 \\ 1 \end{bmatrix}, \quad \mathbf{x}_2 = \begin{bmatrix} 0 \\ 0 \\ 1 \end{bmatrix} \quad \text{and} \quad \mathbf{x}_3 = \begin{bmatrix} 1 \\ -2 \\ 1 \end{bmatrix}.$$

Let

$$Q = \begin{bmatrix} 1 & 0 & 1 \\ 0 & 0 & -2 \\ 0 & 1 & 1 \end{bmatrix}$$

so that

$$Q^{-1} = \frac{1}{2}\begin{bmatrix} 2 & 1 & 0 \\ 0 & 1 & 2 \\ 0 & -1 & 0 \end{bmatrix}.$$

Hence (see [B7] Section 6.4)

$$A^n = QD^nQ^{-1}$$

where

$$D = \text{diag}\left\{1, \ 1, \ \frac{1}{2}\right\}.$$

So

$$A^n = \begin{bmatrix} 1 & \frac{1}{2} - \left(\frac{1}{2}\right)^{n+1} & 0 \\ 0 & \left(\frac{1}{2}\right)^n & 0 \\ 0 & \frac{1}{2} - \left(\frac{1}{2}\right)^{n+1} & 1 \end{bmatrix}$$

and

$$p_n = p_0 + \left[\frac{1}{2} - \left(\frac{1}{2}\right)^{n+1}\right] q_0$$

$$q_n = \left(\frac{1}{2}\right)^n q_0$$

$$r_n = \left[\frac{1}{2} - \left(\frac{1}{2}\right)^{n+1}\right] q_0 + r_0 \qquad (n = 1, 2, \ldots)$$

So for large n

$$p_n \to p_0 + \frac{1}{2} q_0$$

$$q_n \to 0$$

$$r_n \to \frac{1}{2} q_0 + r_0.$$

This interesting result shows that this inbreeding process eventually results in a population with individuals of genotypes AA and aa only.

Model 3

Another inbreeding example concerns a large colony of mice whose colour is determined by their genotype AA, Aa and aa. The breeding programme consists of allowing mating only with a mouse of the colour corresponding to genotype aa. Let p_n, q_n and r_n be proportions of the mice population in the nth generation corresponding to genotypes AA, Aa and aa respectively, so that

$$p_n + q_n + r_n = 1 \qquad (n = 0, 1, 2, \ldots)$$

p_0, q_0, r_0 being the initial distribution of the proportions.

We are interested in the distribution of the proportions

(1) in the fifth generation
(2) in the limit as the number of generation increases indefinitely.

We again make use of Table 1. For example, to obtain the proportion of genotype aa from the preceding distributions we look along the first row and columns 1, 2 and 3 (the only columns involving at least one parent of genotype aa), to obtain

$$r_n = r_{n-1} + \frac{1}{2} q_{n-1}.$$

Similarly

$$q_n = \frac{1}{2} q_{n-1} + p_{n-1}$$

$$p_n = 0$$

or

$$\mathbf{x}^{(n)} = A\,\mathbf{x}^{(n-1)} \qquad n = 1, 2, \ldots$$

where

$$A = \begin{bmatrix} 0 & 0 & 0 \\ 1 & \frac{1}{2} & 0 \\ 0 & \frac{1}{2} & 1 \end{bmatrix}$$

and

$$\mathbf{x}^{(k)} = [p_k, q_k, r_k]' \qquad k = 0, 1, 2, \ldots$$

Hence

$$\mathbf{x}^{(n)} = A^n \mathbf{x}^{(0)}.$$

Again we have a nonnegative matrix A which is reducible so that Theorem 4.9 applies.

The eigenvalues of A are

$$1, \frac{1}{2}, 0$$

and the corresponding eigenvectors are

$$\mathbf{x}_1 = [0, 0, 1]', \quad \mathbf{x}_2 = [0, -1, 1]' \quad \text{and} \quad \mathbf{x}_3 = [1, -2, 1]'.$$

Let

$$Q = \begin{bmatrix} 0 & 0 & 1 \\ 0 & -1 & -2 \\ 1 & 1 & 1 \end{bmatrix}$$

then

$$Q^{-1} = \begin{bmatrix} 1 & 1 & 1 \\ -2 & -1 & 0 \\ 1 & 0 & 0 \end{bmatrix}$$

and

$$D = \operatorname{diag}\left\{1, \frac{1}{2}, 0\right\}$$

Then

$$A^n = QD^nQ^{-1} = \begin{bmatrix} 0 & 0 & 0 \\ 2\left(\frac{1}{2}\right)^n & \left(\frac{1}{2}\right)^n & 0 \\ 1-2\left(\frac{1}{2}\right)^n & 1-\left(\frac{1}{2}\right)^n & 1 \end{bmatrix}.$$

It follows that

$$p_n = 0$$

$$q_n = \left(\frac{1}{2}\right)^{n-1} p_0 + \left(\frac{1}{2}\right)^n q_0$$

$$r_n = \left[1 - \left(\frac{1}{2}\right)^{n-1}\right] p_0 + \left[1 - \left(\frac{1}{2}\right)^n\right] q_0 + r_0$$

for $n = 1, 2, \ldots$

(1) When $n = 5$

$$p_5 = 0$$

$$q_5 = \frac{1}{16} p_0 + \frac{1}{32} q_0$$

$$r_5 = \frac{15}{16} p_0 + \frac{31}{32} q_0 + r_0$$

Using the fact that

$$p_0 + q_0 + r_0 = 1$$

we can write the above results as

$$p_5 = 0$$

$$q_5 = \frac{1}{16} p_0 + \frac{1}{32} q_0$$

$$r_5 = 1 - \frac{1}{16} p_0 - \frac{1}{32} q_0$$

(2) As $n \to \infty$, we obtain

$$p_n = 0$$

$$q_n \to 0$$

$$r_n \to p_0 + q_0 + r_0 = 1$$

Model 4

This model involves various abnormalities and genetic traits which are transmitted from one generation to subsequent ones throught the sex chromosomes. Genes carried on the X-chromosome are called *X-linked*, whereas genes carried on the Y-chromosomes are called *Y-linked*.

The transmission of such genetic traits is known as 'sex-linked inheritance'.

Y-linked inheritance only involves males and is passed on from an affected male to all his sons but to none of his daughters. Abnormalities of these types are extremely rare.

If his X-chromosome is affected the X-linked inheritance is always manifest in the male, but only if both the chromosomes are affected is it manifest in the female. The reason for this is that in many cases the affected (mutant) gene is recessive. Since the female with only one recessive gene has her genotype determined by the dominant (healthy) gene, she will not manifest the abnormality, but she will be a *carrier* and pass on her affected gene to half of her offsprings (on average). The male has only one X-chromosome and if it is affected there is no normal X-chromosome to counteract its affect as in the female. This mutant gene is passed on to all his female offsprings but (obviously) to none of his male ones.

Examples of X-linked recessive traits in man is a muscular disease called Duchenne muscular dystrophy, another is partial colour blindness and a third is haemophilia.

If the haemophilia gene designated by $X^{(h)}$, we can represent some of the various genetic combinations explained above by Figs 9–11.

Female \ Male	X	Y
$X^{(h)}$	$X^{(h)}X$ carrier daughter	$X^{(h)}Y$ affected son
X	XX normal daughter	XY normal son

Fig. 9. Carrier female – normal male.

Female \ Male	$X^{(h)}$	Y
X	$X^{(h)}X$ carrier daughter	XY normal son
X	$X^{(h)}X$ carrier daughter	XY normal son

Fig. 10. Normal female – affected male.

Female \ Male	$X^{(h)}$	Y
$X^{(h)}$	$X^{(h)}X^{(h)}$ affected daughter	$X^{(h)}Y$ affected son
X	$XX^{(h)}$ carrier daughter	XY normal son

Fig. 11. Carrier female – affected male.

In man this case is extremely rare and until fairly recently it was not thought that a woman haemophiliac exists. Apparently a few such cases have been reported recently.

Other tables are easily constructed. For example for affected female–normal male.

To use the previous notation for X-linked inheritance we can denote the female genotypes by

$$\text{AA, Aa, aa (instead of } XX,\ XX^{(h)},\ X^{(h)}X^{(h)})$$

where a now denotes the recessive affected gene.

The male genotypes can now be denoted by

A or a

since only one X-chromosome is involved.

Although the above results are of great interest, particularly in Genetic counselling, it is within the animal and plant kingdoms where inbreeding is the norm, that practical steps can be taken either to eliminate (or to encourage) certain traits.

We consider a model involving X-linked inheritance in which an offspring of each sex is selected at random from parents of given genotypes. These offsprings are mated and the process repeated. The object is to determine the probability of genotypes of the sibling-pairs either for a given generation or in the limit, as the number of generations increase indefinitely.

Let us consider a pair to be of genotypes

(a, AA)

in the $(n-1)$st generation.

Their offsprings (in the nth generation) correspond to the ones given in Fig. 10, since the male is of genotype a (affected), whereas the female is of genotype AA (normal).

The offsprings are therefore:

Male A (normal) and female Aa (carrier).

It follows that the genotypes of the offsprings selected for mating are

(A, Aa)

with probability 1.

Conversely, if the genotypes mated in the nth generation are

(A, Aa)

then the probability that they are the offsprings of genotypes

(a, AA)

in the $(n-1)$st generation is 1.

Next consider that the genotypes of the mating-pair in the $(n-1)$st generation to be

(A, Aa).

Their offsprings (in the nth generation) are a and A (male), AA and Aa (female) (see Fig. 9).

So the possible selections in this case are the pairs

(A, AA), (A, Aa), (a, AA), (a, Aa)

all equally likely, hence each with probability 1/4. Conversely, if the genotypes mated in the nth generation are

(A, AA) [or (A, Aa) or (a, AA) or (a, Aa)]

then the probability that they are the offsprings of genotypes

(A, Aa)

in the $(n-1)$st generation is 1/4.

We can carry out a similar exercise for all possible combinations of genotype pairs and collect the results in a matrix form as

$$\mathbf{x}^{(n)} = A\mathbf{x}^{(n-1)}$$

relating two consecutive generations.

We again obtain

$$\mathbf{x}^{(n)} = \mathbf{A}^n\mathbf{x}^{(0)} \qquad (n = 1, 2, \ldots).$$

Let

p_n = probability that the genotypes in the nth generation are (A, AA).

q_n = probability that the genotypes in the nth generation are (a, aa).

r_n = probability that the genotypes in the nth generation are (a, AA).

s_n = probability that the genotypes in the nth generation are (A, Aa).

t_n = probability that the genotypes in the nth generation are (a, Aa).

u_n = probability that the genotypes in the nth generation are (A, aa).

$\mathbf{x}^{(n)} = [p_n, q_n, r_n, s_n, t_n, u_n]' \qquad (n = 0, 1, 2, \ldots)$

Then

$$A = \left[\begin{array}{cc:cccc} 1 & 0 & 0 & \frac{1}{4} & 0 & 0 \\ 0 & 1 & 0 & 0 & \frac{1}{4} & 0 \\ \hdashline 0 & 0 & 0 & \frac{1}{4} & 0 & 0 \\ 0 & 0 & 1 & \frac{1}{4} & \frac{1}{4} & 0 \\ 0 & 0 & 0 & \frac{1}{4} & \frac{1}{4} & 1 \\ 0 & 0 & 0 & 0 & \frac{1}{4} & 0 \end{array}\right].$$

This is a nonnegative matrix (indeed A is column stochastic) so that Theorem 4.9 applies. Unfortunately, as can be seen when A is considered in a partitioned form, A has at least two eigenvalues of unit modulus, so that the methods discussed in Chapter 6 for evaluating A^n are not directly applicable.

On the other hand, to find the eigenvalues of A we need only evaluate the eigenvalues of

$$A_1 = \begin{bmatrix} 0 & \frac{1}{4} & 0 & 0 \\ 1 & \frac{1}{4} & \frac{1}{4} & 0 \\ 0 & \frac{1}{4} & \frac{1}{4} & 1 \\ 0 & 0 & \frac{1}{4} & 0 \end{bmatrix}.$$

Since the characteristic equation of A_1 is

$$\left(\lambda^2 - \frac{1}{4}\right)\left(\lambda^2 - \frac{1}{2}\lambda - \frac{1}{4}\right) = 0$$

the eigenvalues of A are

$$\lambda_1 = 1,\ \lambda_2 = 1,\ \lambda_3 = \frac{1}{2},\ \lambda_4 = -\frac{1}{2},\ \lambda_5 = \frac{1}{4}(1+\sqrt{5}),$$

$$\lambda_6 = \frac{1}{4}(1-\sqrt{5})$$

the corresponding eigenvectors are:

$$\mathbf{y}_1 = \begin{bmatrix} 1 \\ 0 \\ 0 \\ 0 \\ 0 \\ 0 \end{bmatrix} \quad \mathbf{y}_2 = \begin{bmatrix} 0 \\ 1 \\ 0 \\ 0 \\ 0 \\ 0 \end{bmatrix} \quad \mathbf{y}_3 = \begin{bmatrix} 1 \\ -1 \\ -1 \\ -2 \\ 2 \\ 1 \end{bmatrix} \quad \mathbf{y}_4 = \begin{bmatrix} 1 \\ -1 \\ 3 \\ -6 \\ 6 \\ -3 \end{bmatrix}$$

$$\mathbf{y}_5 = \begin{bmatrix} 1 \\ 1 \\ 2-\sqrt{5} \\ -3+\sqrt{5} \\ -3+\sqrt{5} \\ 2-\sqrt{5} \end{bmatrix} \quad \mathbf{y}_6 = \begin{bmatrix} -1 \\ -1 \\ -2-\sqrt{5} \\ 3+\sqrt{5} \\ 3+\sqrt{5} \\ -2-\sqrt{5} \end{bmatrix}.$$

Let

$$Q = \begin{bmatrix} 1 & 0 & 1 & 1 & 1 & -1 \\ 0 & 1 & -1 & -1 & 1 & -1 \\ 0 & 0 & -1 & 3 & 2-\sqrt{5} & -2-\sqrt{5} \\ 0 & 0 & -2 & -6 & -3+\sqrt{5} & 3+\sqrt{5} \\ 0 & 0 & 2 & 6 & -3+\sqrt{5} & 3+\sqrt{5} \\ 0 & 0 & 1 & -3 & 2-\sqrt{5} & -2-\sqrt{5} \end{bmatrix}.$$

Writing Q in a partitioned form as

$$Q = \left[\begin{array}{c:c} I & A_1 \\ \hdashline O & A_2 \end{array}\right]$$

we obtain

$$Q^{-1} = \left[\begin{array}{c:c} I & B_1 \\ \hdashline O & B_2 \end{array}\right]$$

where (as easily verified)

$$B_2 = A_2^{-1} \quad \text{and} \quad B_1 = -A_1 A_2^{-1}.$$

We find

$$Q^{-1} = \begin{bmatrix} 1 & 0 & \frac{2}{3} & \frac{2}{3} & \frac{1}{3} & \frac{1}{3} \\ 0 & 1 & \frac{1}{3} & \frac{1}{3} & \frac{2}{3} & \frac{2}{3} \\ 0 & 0 & -\frac{1}{4} & -\frac{1}{8} & \frac{1}{8} & \frac{1}{4} \\ 0 & 0 & \frac{1}{12} & -\frac{1}{24} & \frac{1}{24} & -\frac{1}{12} \\ 0 & 0 & -\frac{3+\sqrt{5}}{4\sqrt{5}} & -\frac{2+\sqrt{5}}{4\sqrt{5}} & -\frac{2+\sqrt{5}}{4\sqrt{5}} & -\frac{3+\sqrt{5}}{4\sqrt{5}} \\ 0 & 0 & -\frac{3-\sqrt{5}}{4\sqrt{5}} & -\frac{2-\sqrt{5}}{4\sqrt{5}} & -\frac{2-\sqrt{5}}{4\sqrt{5}} & -\frac{3-\sqrt{5}}{4\sqrt{5}} \end{bmatrix}.$$

Also

$$D^n = \text{diag}\left\{1,\ 1,\ \left(\frac{1}{2}\right)^n,\ \left(-\frac{1}{2}\right)^n,\ \left(\frac{1+\sqrt{5}}{4}\right)^n,\ \left(\frac{1-\sqrt{5}}{4}\right)^n\right\}.$$

For example for $n = 5$

$$D^5 = \{1,\ 1,\ 0.031,\ -0.031, 0.347,\ -0.003\}.$$

Choosing

$$\mathbf{x}^{(0)} = [0,\ 0,\ 0,\ 0,\ 1,\ 0]'$$

so that the original pair have genotypes (a, Aa) we find in the fifth generation

$$\mathbf{x}^{(5)} = [0.17,\ 0.5,\ 0.04,\ 0.12,\ 0.12,\ 0.04]'$$

(approximate solution).

This implies that in the fifth generation the probabilities that the selected pairs are of genotype

(A, AA), (a, aa), (a, AA), (A, Aa), (a, Aa), (A, aa)

are respectively

0.17, 0.5, 0.04, 0.12, 0.12, 0.04.

As the number of generations increases indefinitely $n \to \infty$, so that

$$D^n \to \text{diag}\{1,\ 1,\ 0,\ 0,\ 0,\ 0\}.$$

Then, starting with the previously defined $\mathbf{x}^{(0)}$, we find

$$\mathbf{x}^{(n)} \to \left[\frac{1}{3}, \frac{2}{3}, 0, 0, 0, 0\right]'.$$

7.3 SOME ECONOMIC MODELS

The Leontief Models

As a result of analysing US economic performance for the periods 1919 and 1929, Wassily Leontief published his ideas on input–output models in 1942 and 1951.

We consider two models; the *closed* model and the *open* model.

Model 1

Leontief's Closed (input–output) Model

This model considers an economic system consisting of a finite number n of industries each producing only one commodity, so that there is no joint production.

Each industry produces an output (a good or a service) which is just sufficient to meet the demands of the n industries, in other words the model assumes constant proportions between the inputs and the ouput for each industry.

We use the following notation:

Y_i = gross output of industry $i = 1, 2, \ldots, n$.

y_{ij} = the amount of the ith commodty needed as input by the jth industry.

We define the *input coefficients*

$$a_{ij} = \frac{y_{ij}}{Y_j} \qquad (i, j = 1, 2, \ldots, n) \tag{7.1}$$

it is the amount of the ith commodity needed to produce *one unit* of the jth commodity.

Notice that

$$a_{ij} \geqslant 0 \qquad (i, j = 1, 2, \ldots, n)$$

The nonnegative matrix

$$A = [a_{ij}] \tag{7.2}$$

is then such

$$a_{1j} + a_{2j} + \ldots + a_{nj} = 1 \qquad (j = 1, 2, \ldots, n)$$

since each column of A specifies the input requirements for the production of *one unit* of output of a particular commodity.

The 'correct' equilibrium for this model is obtained by considering the ouput of the ith industry ($i = 1, 2, \ldots, n$) being equal to the required inputs of the ith commodity to the n industries, so that

$$\sum_{j=1}^{n} y_{ij} = Y_i \qquad i = 1, 2, \ldots, n \tag{7.3}$$

From (7.1), we have

$$y_{ij} = a_{ij} Y_j \tag{7.4}$$

so that O substituting (7.4) into (7.3), we obtain

$$\sum_{j=1}^{n} a_{ij} Y_j = Y_i > 0 \qquad (i = 1, 2, \ldots, n)$$

which can be written in matrix form as,

$$A\mathbf{Y} = \mathbf{Y} \tag{7.5}$$

where

$$\mathbf{Y} = [Y_1, Y_2, \ldots, Y_n]' > \mathbf{0}$$

So the problem in this situation is to find the necessary condition for the existence of a $\mathbf{Y} > \mathbf{0}$ which satisfies the equilibrium condition (7.5).

A similar formula to (7.5) above is obviously obtained when considering the prices of the commodities involved.

Indeed, if

X_i = value of gross output of industry i; ($i = 1, 2, \ldots, n$)

x_{ij} = value of the output from the ith industry used as input to industry j ($i, j = 1, 2, \ldots, n$).

and

$$a_{ij} = \frac{x_{ij}}{X_j}$$

Then

$$A\mathbf{X} = \mathbf{X}$$

or

$$[I - A]\mathbf{X} = \mathbf{0} \tag{7.6}$$

where

$$\mathbf{X} = [X_1, X_2, \ldots, X_n]'.$$

Example 7.1

A region has three industries; a steel plant, an electricity generating plant and a coalmine.

To mine £1000 of coal, the mining company uses £200 of electricity, £60 of steel and £200 of its own coal. To manufacture £1000 of steel the plant buys £500 of electricity, £100 of coal and £400 of its own products. To generate £1000 of electricity, the plant uses £700 of coal, £200 of steel products and £100 of electricity.

Find a linearly independent $\mathbf{X} > \mathbf{0}$ to satisfy the equilibrium equation (7.6).

Solution
For this example

$$A = \begin{bmatrix} 0.2 & 0.1 & 0.7 \\ 0.6 & 0.4 & 0.2 \\ 0.2 & 0.5 & 0.1 \end{bmatrix}.$$

Since the column sums of A are all equal to one,

$$r = \rho(A) = 1$$

(see Section 4.3).

It follows that the matrix

$$[I - A]$$

is singular, so that (7.6) has a non-trivial solution.

Furthermore, by the Perron–Frobenius theorem for positive matrices (Section 4.3), the eigenvector corresponding to $r = 1$ is such that

$$A\mathbf{x} = 1\mathbf{x}$$

where

$$\mathbf{x} > \mathbf{0}.$$

So this eigenvector satisfies the equilibrium condition (7.6).

Indeed it is simple to calculate $\mathbf{x}$, it is

$$\mathbf{x} = \left[1, \frac{29}{22}, \frac{21}{22}\right]'. \qquad \blacksquare$$

In the above example we have considered all the entries a_{ij} of A to be positive. In general the entries a_{ij} are nonnegative, so the question arises under what conditions does an unique equilibrium condition $\mathbf{X}$ exist and is positive (or *viable* solution).

The answer is very simple when taking account of the Perron–Frobenius theorem for irreducible matrices (Section 4.4). This shows that so long as A is irreducible and $\rho(A) = 1$, (7.6) has a solution $\mathbf{X}$ which is positive.

(*Note*: when requiring an unique solution, we are referring to a linearly independent solution, that is a solution unique up to a scalar multiple.)

Model 2

Leontief's Open Model

This is an input–output model of an economy consisting of a sector of n industries each producing one commodity (as for the closed model).

The inputs into the economy consists of the n commodities together with one (or possibly more) inputs which is not an output of the production process. The most frequently used example of such an input is the labour force needed to run the economy. Generally there is an excess in production to satisfy an outside demand.

We use the notation defined for the closed model. For example

$$x_i = \text{value of the total output of industry } i.$$

Since in contrast with the closed model, the open model must satisfy one (or possibly more) outside demand, the column sums of the entries of the corresponding matrix $A = [a_{ij}]$ will be (in general) less than 1.

In other words we now have

$$a_{1j} + a_{2j} + \ldots + a_{nj} \leqslant 1 \qquad (j = 1, 2, \ldots, n)$$

where $a_{ij} \geqslant 0$ (all i, j).

Let

$$\mathbf{X} = [X_1, X_2, \ldots, X_n]$$

be the vector of the value of total outputs of the n industries, then $A\mathbf{X}$ is the vector of values of the inputs required for the n industries to produce those outputs.

It follows that

$$\mathbf{X} - A\mathbf{X}$$

in the vector of excess of outputs over the inputs which are necessary to manufacture these outputs.

Let $\mathbf{C}$ = a vector representing an arbitrary set of outputs. $\mathbf{C}$ is often called a *final demand* or the *bill of goods*.

The problem for the open model is to find the necessary conditions for the existence of a vector $\mathbf{x} \geqslant \mathbf{0}$ such that

$$\mathbf{x} - A\mathbf{x} = \mathbf{C}$$

or equivalently

$$[I - A]\mathbf{x} = \mathbf{C}. \tag{7.7}$$

Example 7.2

Over the years, new technology having been introduced, the economic situation has changed in the region considered in Example 7.1.

To mine £1000 of coal, the company now uses £350 of electricity and £150 of steel. To manufacture £1000 of steel the plant uses £500 of coal, £100 of electricity and £100 of steel. To generate £1000 of electricity the plant uses £600 of coal and £100 of electricity.

This year the region has received an outside order for £250 000 worth of coal and £500 000 of steel.

Determine how much each industry must produce to exactly meet the demands both from the region and from the outside.

Solution

$$A = \begin{bmatrix} 0 & 0.5 & 0.6 \\ 0.15 & 0.1 & 0 \\ 0.35 & 0.1 & 0.1 \end{bmatrix}$$

hence

$$I - A = \begin{bmatrix} 1 & -0.5 & -0.6 \\ -0.15 & 0.9 & 0 \\ -0.35 & -0.1 & 0.9 \end{bmatrix}$$

and

$$[I-A]^{-1} = \frac{1}{0.5445} \begin{bmatrix} 0.81 & 0.51 & 0.54 \\ 0.135 & 0.69 & 0.09 \\ 0.33 & 0.275 & 0.825 \end{bmatrix}.$$

Since

$$\mathbf{C} = [250\,000, \quad 500\,000, \quad 0]'$$

it follows that

$$\mathbf{x} = [I-A]^{-1}\,\mathbf{C} = [840\,220, \quad 378\,750, \quad 220\,000]'.$$

This means that to meet all demands the output of coal should be £840 220, the output of steel should be £378 750 and the output of the power generated should be £220 000. ■

In the above example we note that when $\mathbf{C} \geqslant \mathbf{0}$, for a solution $\mathbf{X} > \mathbf{0}$, $[I-A]$ must be non-singular and $[I-A]^{-1}$ should be positive.

By Theorem 4.5, we know that $[I-A]$ is non-singular, since for the Open Model the columns sums of A cannot all be unity (at least one column sum is less than 1), hence

$$\rho(A) < 1$$

So $[I-A]$ is a M-matrix (see Chapter 5). By Lemma 5.1, $[I-A]^{-1}$ is always nonnegative. Furthermore, if A is irreducible, this guarantees that $[I-A]^{-1}$ is positive (see note to Lemma 5.1).

7.4 SOME MARKOV CHAIN MODELS

Model 1

We consider a molecule which at time points $t_1, t_2, t_3, \ldots$ can be at one of two possible energy levels E_1 and E_2 (say).

Given that the molecule is at E_1 at some time point, the probability that it moves to E_2 at the subsequent time point is a. If it is at E_2, the probability that it moves to E_1 at the subsequent time point is b.

This gives rise to a Markov chain with states S_0 and S_1 (say) where (for example)

S_0 stands for 'no molecule at energy level E_1'

S_1 stands for 'one molecule at energy level E_1'

The problem is to determine the distribution of this process after a period of n time points, given an initial distribution.

What is the distribution as $n \to \infty$?

To study this rather simple process we consider the associated transition matrix, it is

$$P = \begin{bmatrix} 1-a & a \\ b & 1-b \end{bmatrix}.$$

The eigenvalues of P are

$$\lambda_1 = 1 \quad \text{and} \quad \lambda_2 = 1 - a - b.$$

The associated (right) eigenvectors are

$$\mathbf{x}_1 = [1,\ 1]' \quad \text{and} \quad \mathbf{x}_2 = \left[\frac{a}{1-\lambda_2},\ -\frac{b}{1-\lambda_2}\right]'$$

where it is assumed that

$$a > 0,\ b > 0 \quad \text{and} \quad a + b \neq 1.$$

Let

$$Q = \begin{bmatrix} 1 & \dfrac{a}{1-\lambda_2} \\ 1 & -\dfrac{b}{1-\lambda_2} \end{bmatrix},$$

then

$$Q^{-1} = \begin{bmatrix} \dfrac{b}{1-\lambda_2} & \dfrac{a}{1-\lambda_2} \\ 1 & -1 \end{bmatrix}$$

and

$$P^n = Q\Lambda^n Q^{-1}$$

where

$$\Lambda = \text{diag}\,\{1,\ 1-a-b\} = \text{diag}\,\{1,\ \lambda_2\}.$$

It follows that

$$P^n = \frac{1}{a+b}\begin{bmatrix} b & a \\ b & a \end{bmatrix} + \frac{(1-a-b)^n}{a+b}\begin{bmatrix} a & -a \\ -b & b \end{bmatrix}.$$

The initial distribution is of the form [1, 0] or [0, 1] so that it is now very simple to obtain the distribution for any specified n, given an initial distribution.

Since it is assumed that

$$|1-a-b| < 1.$$

It follows that as $n \to \infty$

$$P^n \to \frac{1}{a+b}\begin{bmatrix} b & a \\ b & a \end{bmatrix}$$

which shows that whatever the initial distribution, the process tends to a *stationary* distribution

$$\frac{1}{a+b}\ [b, a]'$$

so that we are dealing with an ergodic system (see Definition 6.5). Notice that we can obtain the last result in a more elegant manner by exploiting the theory developed in Chapter 6 (see Example 6.1).

It was shown there that

$$\lim_{n\to\infty} P^n = \mathbf{e}\mathbf{z}'$$

where

$$\mathbf{z}' = (\mathbf{y}'\mathbf{e})^{-1}\mathbf{y}'$$

where $\mathbf{y}$ is the eigenvector of P' corresponding to $\lambda = 1$. Indeed, for the above matrix P

$$\mathbf{y}' = \left[\frac{b}{a+b},\ \frac{a}{a+b}\right]$$

It follows that $\mathbf{y}'\mathbf{e} = 1$, so that

$$\mathbf{z}' = \mathbf{y}'$$

and

$$\lim_{n\to\infty} P^n = \begin{bmatrix} 1 \\ 1 \end{bmatrix} \begin{bmatrix} \dfrac{b}{a+b} & \dfrac{a}{a+b} \end{bmatrix} = \frac{1}{a+b} \begin{bmatrix} b & a \\ b & a \end{bmatrix}.$$

There are two extreme cases for this model.

(1) $a = b = 0$.
In this case

$$P = \begin{bmatrix} 1 & 0 \\ 0 & 1 \end{bmatrix}$$

so that both states are absorbing. No transition of the molecule is possible between S_0 and S_1.

(2) $a = b = 1$.
In this case

$$P = \begin{bmatrix} 0 & 1 \\ 1 & 0 \end{bmatrix}$$

This is a cyclic chain, in which transition is certain. The molecule alternates between states S_0 and S_1 at successive time points.

This model can be generalised to involve any number N of molecules, each moving independently between the two energy levels. If, for example, we consider X_n to the number of molecules at E_1 (at time n), then

$$\{X_n, \ n \geqslant 0\}$$

is a Markov chain with $(N + 1)$ states.

Model 2

In this model we analyse a typical Queueing System.

A harbour designed to service cargo ships has the following characteristics.

(1) It can accommodate two ships only. If a ship arrives and there are already two ships in the harbour, it is sent away.
(2) The arrival and departure of ships can only occur when tides allow, that is at discrete time points $t_1, t_2, t_3, \ldots$ (assumed equally spaced).
(3) Only one ship can be serviced at a time, during one time period.
(4) The entrance is just large enough to allow one ship arrival and one ship departure only at each time point.
(5) At each time point there is a probability p_1 that a ship arrives to be serviced.
(6) At each time point there is a probability p_2 that a ship is served and departs (that assumes that there is a ship in the harbour to be served).
(7) The arrival and departure of ships is assumed to be statistically independent.

As usual in this type of problem we are interested in finding the probabilities of 0, 1 or 2 ships in the queue waiting to be serviced after a specified number n of time points (or steps in the Markov chain terminology). Also what is the queue distribution as n increases indefinitely?

To set up the transition matrix for this model, we define the states of the system as:

S_i = the number i of ships in the harbour at any time point ($i = 0, 1, 2, \ldots$).
Let

$$q_1 = 1 - p_1,$$

$$q_2 = 1 - p_2$$

and

$$P = [p_{ij}]$$

be the transition matrix.

(Do not confuse the elements p_{ij} of the transition matrix with the probabilities p_1 and p_2.)

For example,

p_{12} is the probability that the system changes at a time point from the state '1 ship in the harbour' to the state '2 ships in the harbour'.

This event can only occur if

1 ship arrives
no ship leaves the harbour.

Since those events are mutually exclusive, we have

$$p_{12} = p_1 q_2.$$

Similarly

P_{22} is the probability that the system changes from '2 ships in the harbour' to '2 ships in the harbour'.

This occurs if

no ship arrives and no ship leaves or
one ship arrives and one ship leaves or
one ship arrives and no ship leaves.

Note: The last event occurs when a ship arrives at the harbour and is sent away since two ships are already there (see (1) above). So no ship leaves the harbour.

Since the events are mutally exclusive

$$p_{22} = q_1 q_2 + p_1 p_2 + p_1 q_2 = p_1 p_2 + q_2.$$

Calculating all the p_{ij} $(i, j = 0, 1, 2)$ in this manner, we finally obtain

$$P = \begin{bmatrix} q_1 & p_1 & 0 \\ q_1 p_2 & p_1 p_2 + q_1 q_2 & p_1 q_2 \\ 0 & p_2 q_1 & p_1 p_2 + q_2 \end{bmatrix}.$$

For example, if

$$p_1 = \frac{1}{3} \quad \text{and} \quad p_2 = \frac{1}{4},$$

then

$$P = \begin{bmatrix} \frac{2}{3} & \frac{1}{3} & 0 \\ \frac{1}{6} & \frac{1}{12} & \frac{1}{4} \\ 0 & \frac{1}{1} & \frac{5}{6} \end{bmatrix}.$$

The eigenvalues of P are

$$\lambda_1 = 1, \; \lambda_2 = \frac{3}{4} \quad \text{and} \quad \lambda_3 = \frac{1}{3}.$$

and the corresponding eigenvectors

$$\mathbf{x}_1 = \begin{bmatrix} 1 \\ 1 \\ 1 \end{bmatrix}, \quad \mathbf{x}_2 = \begin{bmatrix} 4 \\ 1 \\ -2 \end{bmatrix}, \quad \text{and} \quad \mathbf{x}_3 = \begin{bmatrix} 1 \\ -3 \\ 1 \end{bmatrix}.$$

(Since P is a stochastic matrix, we know λ_1 and $\mathbf{x}_1$ and that all other eigenvalues λ_i are such that $|\lambda_i| \leqslant 1$.)

Let

$$Q = \begin{bmatrix} 1 & 4 & 3 \\ 1 & 1 & -3 \\ 1 & -2 & 1 \end{bmatrix}$$

then

$$Q^{-1} = \frac{1}{30} \begin{bmatrix} 5 & 10 & 15 \\ 4 & 2 & -6 \\ 3 & -6 & 3 \end{bmatrix}$$

and

$$\Lambda = \text{diag}\left\{1, \frac{3}{4}, \frac{1}{3}\right\}.$$

$$P^n = Q\Lambda^n Q^{-1} = \begin{bmatrix} \frac{1}{6} & \frac{1}{3} & \frac{1}{2} \\ \frac{1}{6} & \frac{1}{3} & \frac{1}{2} \\ \frac{1}{6} & \frac{1}{3} & \frac{1}{2} \end{bmatrix} + \left(\frac{3}{4}\right)^n \begin{bmatrix} \frac{8}{15} & \frac{4}{15} & -\frac{12}{15} \\ \frac{2}{15} & \frac{1}{15} & -\frac{3}{15} \\ -\frac{4}{15} & -\frac{2}{15} & \frac{6}{15} \end{bmatrix}$$

$$+ \left(\frac{1}{3}\right)^n \begin{bmatrix} \frac{3}{10} & -\frac{6}{10} & \frac{3}{10} \\ -\frac{3}{10} & \frac{6}{10} & -\frac{3}{10} \\ \frac{1}{10} & -\frac{2}{10} & \frac{1}{10} \end{bmatrix}.$$

The possible initial states are

$$[1 \ 0 \ 0], \ [0 \ 1 \ 0] \quad \text{or} \quad [0 \ 0 \ 1]$$

depending on whether initially there is 0, 1 or 2 ships in the harbour.

Assuming the initial state to be $[1 \ 0 \ 0]$, we can now obtain the probabilities that there are 0, 1 or 2 ships in the harbour at $t = t_4$ (say), that is when $n = 3$.

From above

$$P^3 = \begin{bmatrix} 0.404 & 0.423 & 0.173 \\ 0.212 & 0.384 & 0.404 \\ 0.058 & 0.270 & 0.672 \end{bmatrix}.$$

So

$$[1 \ 0 \ 0] \, P^3 = [0.404 \ 0.423 \ 0.173].$$

So there is no ship in harbour with probability 0.404, one ship with probability 0.423 and two ships with probability 0.173.

It is also clear that as $n \to \infty$

$$\lim_{n\to\infty} P^n = \begin{bmatrix} \frac{1}{6} & \frac{1}{3} & \frac{1}{2} \\ \frac{1}{6} & \frac{1}{3} & \frac{1}{2} \\ \frac{1}{6} & \frac{1}{3} & \frac{1}{2} \end{bmatrix}$$

Since we are dealing with a non-cyclic ergodic (i.e. regular) Markov chain, we can obtain this last result by the technique of Example 6.1.

Indeed, the eigenvector of P' corresponding to $\lambda = 1$, is

$$\mathbf{y}' = [1 \;\; 2 \;\; 3]$$

so that

$$\mathbf{z}' = \frac{1}{6}\,[1 \;\; 2 \;\; 3]\,,$$

and

$$\lim_{n\to\infty} P^n = \frac{1}{6}\begin{bmatrix} 1 \\ 1 \\ 1 \end{bmatrix} [1 \;\; 2 \;\; 3] = \frac{1}{6}\begin{bmatrix} 1 & 2 & 3 \\ 1 & 2 & 3 \\ 1 & 2 & 3 \end{bmatrix}.$$

This implies that in the long term the process tends to a stationary distribution, the probabilities of 0, 1, 2 ships in harbour are

$$\frac{1}{6},\; \frac{1}{3},\; \frac{1}{2}$$

respectively, whatever the initial condition.

Model 3

In 1907 Ehrenfest proposed a model to explain, in terms of molecular displacement, the exchange of heat between two isolated bodies at different temperatures.

The model consists of two urns A and B (say) containing a total of N balls labelled $1, 2, \ldots, N$.

At each time point $t_1, t_2, t_3, \ldots$ an integer between 1 and N is selected at random. The corresponding ball is removed from its urn and replaced in the other one. This process is repeated in such a way that the selections are statistically independent.

Let X_n be the number of balls in the urn (say) at t_n, then

$$\{X_n\}$$

is a Markov chain on the set of states $\{0, 1, 2, \ldots, N\}$.

The transition matrix is

$$P = [p_{ij}]$$

where p_{ij} is the probability that the process is in state i (i balls in urn A) at a time point t_i and in state j at the consecutive time point t_{i+1}.

Given that the process is in state i at time t_i, the probability that the random integer selected corresponds to a number on a ball in urn A is

$$\frac{i}{N}.$$

In this case, there will be $(i-1)$ balls in the urn at time point t_{i+1}.

Similarly, the probability that the integer selected corresponds to a number on a ball in urn B is

$$\frac{N-i}{N}$$

in which case there will be $(i+1)$ balls in urn A at t_{i+1}.

So

$$p_{ij} = \begin{cases} \dfrac{i}{N}, & j = i-1 \\ 1 - \dfrac{i}{N} & j = i+1 \\ 0 & \text{otherwise} \end{cases}$$

So for $N = 3$ (for example)

$$P = \begin{bmatrix} 0 & 1 & 0 & 0 \\ \frac{1}{3} & 0 & \frac{2}{3} & 0 \\ 0 & \frac{2}{3} & 0 & \frac{1}{3} \\ 0 & 0 & 1 & 0 \end{bmatrix}.$$

The object of setting up this process is to determine the limiting probability distribution of the number of balls (molecules) in urn A.

Observing that the matrix P represents a regular Markov chain we can once again use the technique of Example 6.1 to determine the limiting distribution.

The eigenvectors of P and P' corresponding to the eigenvalue $\lambda = 1$ are

$$\mathbf{x}' = [1,\ 1,\ 1,\ 1]$$

and

$$y' = [2,\ 6,\ 6,\ 2]$$

respectively.

So

$$\lim_{n\to\infty} P^n = \frac{1}{16}\begin{bmatrix}1\\1\\1\\1\end{bmatrix}[2\ \ 6\ \ 6\ \ 2]$$

$$= \begin{bmatrix} \frac{1}{8} & \frac{3}{8} & \frac{3}{8} & \frac{1}{8} \\ \frac{1}{8} & \frac{3}{8} & \frac{3}{8} & \frac{1}{8} \\ \frac{1}{8} & \frac{3}{8} & \frac{3}{8} & \frac{1}{8} \\ \frac{1}{8} & \frac{3}{8} & \frac{3}{8} & \frac{1}{8} \end{bmatrix}.$$

So in the long run the probabilities of 0, 1, 2, 3 balls in urn A are

$$\frac{1}{8},\ \frac{3}{8},\ \frac{3}{8},\ \frac{1}{8}$$

respectively, whatever the initial distribution.

Appendix

In this appendix we aim to prove the following:

Lemma 1

Let S be a set of positive integers which is closed under addition, so that if $x, y \in S$ then $x + y \in S$, and let d be the greatest common divisor of the integers, then all sufficiently large multiples of $d \in S$.

Before we can prove this lemma, we shall define the terminology used and prove a number of preliminary results.

Let $\mathbf{Z}$ denote the set of integers (positive and negative). If a, b, $q \in \mathbf{Z}$ and

$$aq = b$$

then a is said to be a *divisor* of b and b is said to be a *multiple* of a for some integer q.

If $T = \{t_1, t_2, \ldots, t_n; t_i \in \mathbf{Z}\}$ then any positive integer which is a divisor (or divides) each t_i, is said to be a *common divisor* of T. The largest of these common divisors is called the *greatest common divisor* (g.c.d.) of T.

We shall denote it by d, so that

$$\text{g.c.d.}\,(T) = d.$$

The concept of the g.c.d. can be generalised to any countable set, in particular the set S.

Let

(1) $S_k = \{a_1, a_2, \ldots, a_k; a_r \in S\}$

(2) $S_{k+1} \supseteq S_k \qquad (k = 1, 2, \ldots)$

(3) $\lim_{k \to \infty} S_k = S$ and

(4) $d_k = \text{g.c.d.}\,(S_k)$.

Then we can define the g.c.d. of S as

$$d = \lim_{k \to \infty} d_k.$$

Lemma 2

If $T = \{t_1, t_2, \ldots, t_n; t_i \in \mathbf{Z}\}$ be closed under addition and subtraction. Assume that a least positive integer $m \in T$, so that

$$m = \min\{t_i \in T;\ t_i > 0\},$$

then

$$T = \{q_i m;\ q_i \in \mathbf{Z}\}$$

Proof

We can assume that $T \neq \{0\}$. Let $t \in T$, then

$$t - t = 0 \in T \quad \text{and} \quad 0 - t = -t \in T.$$

It follows that $|t|$ (a positive integer) $\in T$, so that there is a minimum positive integer $m \in T$.

Assume $km \in T$ (k is any positive integer), then

$$(k+1)m = km + m \in T$$

it follows that T contains all positive integral multiples of m.

Also $(-k)m = 0 - km \in T$, so that T contains all negative integral multiples of m.

Let $t \in T$, then there exist $q, r \in \mathbf{Z}$ such that

$$t = qm + r$$

where

$$0 \leqslant r < m$$

so that

$$r = t - qm.$$

But $t \in T$ and $qm \in T$, it follows that

$$r \in T.$$

But m is the *smallest* positive integer in T, hence since $r \geqslant 0$, it follows that

$$r = 0.$$

Since t is any element in T, and

$$t = qm$$

the lemma is proved.

Lemma 3
Let $T = \{t_1, t_2, \ldots, t_m; t_i \in \mathbf{Z}\}$ and g.c.d. $(T) = d$.
Then we can write d in the form

$$d = \sum_{i=1}^{n} a_i t_i$$

where

$$a_i \in \mathbf{Z} \qquad (i = 1, 2, \ldots, n)$$

Proof
Let

$$Q = \left\{q_1, q_2, \ldots, ; q_r = \sum_{i=1}^{n} a_i^{(r)} t_i;\ a_i^{(r)} \in \mathbf{Z}\right\}$$

then

$$q_r \pm q_s = \sum_{i=1}^{n} (a_i^{(r)} \pm a_i^{(s)})\, t_i \in Q$$

since

$$a_i^{(r)} \pm a_i^{(s)} \in \mathbf{Z},$$

So **Q** is closed under addition and subtraction. By Lemma 2,

$$Q = \{p_i m;\ p_i \in \mathbf{Z}\}$$

where

$$m = \sum_{i=1}^{n} a_i t_i > 0.$$

(m is defined in Lemma 2).
But d is a divisor of $t_1, t_2, \ldots, t_n$, hence

d is a divisor of m,

so that

$$m \geqslant d > 0.$$

If we choose

$$a_1^{(1)} = 1 \quad \text{and} \quad a_2^{(1)} = a_3^{(1)} = \ldots = a_n^{(1)} = 0,$$

then

$$q_1^{(1)} = t_1 \in Q$$

Similarly we can show that

$$t_i \in Q \qquad (i = 1, 2, \ldots, n).$$

Hence, by Lemma 2

$$t_i = q_i m, \; q_i \in \mathbf{Z} \qquad (i = 1, 2, \ldots, n)$$

which shows that m is a common divisor of T.

But d is the greatest common divisor of T, hence

$$d \geqslant m > 0.$$

It follows that

$$d = m = \sum_{i=1}^{n} a_i t_i.$$

Lemma 4

Given the set V of positive integers (closed under addition) and g.c.d. $(V) = d$, then there exists a finite subset T of V with

$$\text{g.c.d.}\ (T) = d.$$

Proof

The proof follows without much difficulty from our definition of the g.c.d. of V.

Now we can prove Lemma 1.

Let $V = \{v_1, v_2, \ldots\}$ be a set of positive integers. By dividing each v_i by d, the g.c.d. $(V) = 1$. There is no loss in generality by considering $d = 1$, and to prove the lemma we must now show that all sufficiently large positive integers are in V.

By Lemma 4, there exists a finite subset $T = \{t_1, t_2, \ldots, t_n\}$ of V with

$$\text{g.c.d.}\ (T) = 1.$$

By Lemma 3,

$$1 = \sum_{i=1}^{n} a_i t_i; \; a_i \in \mathbf{Z} \qquad (i = 1, 2, \ldots, n).$$

Summing together the positive terms (P) and the negative terms (N) in the above sum, the above equation can be written as

$$1 = P - N$$

where P and N are positive integers.

Both P and N have the form

$$\sum a_i t_i$$

it follows that $P, N \in V$.

Consider now a positive integer x. If $x > N$, there is a positive integer a such that

$$x = aN + r$$

where

$$0 \leqslant r < N$$

Since

$$P - N = 1,$$

we can write

$$x = aN + r(P - N)$$

$$= (a - r)N + rP.$$

Hence

$$x \in V$$

if

$$a - r \geqslant 0.$$

But $a \geqslant r$ for all r so long as $a \geqslant N - 1$. It follows that $x \in V$ whenever $x \geqslant (N - 1)N$.

Solutions to Problems

CHAPTER 2

2.1 Let $\mathbf{w}_2' = (x, y, z)$ be orthogonal to $\mathbf{u}_1$, then

$$0 = (\mathbf{u}_1, \mathbf{w}_2) = \frac{1}{2}x + \frac{1}{2}y + \frac{\sqrt{2}}{2}z$$

Then

$$\mathbf{w}_2' = (1,\ 1,\ -\sqrt{2})$$

so that

$$\mathbf{u}_2' = \left(\frac{1}{2},\ \frac{1}{2},\ -\frac{1}{\sqrt{2}}\right).$$

Let $\mathbf{w}_3$ be orthogonal to $\mathbf{u}_1$ and $\mathbf{u}_2$, then

$$0 = (\mathbf{u}_1, \mathbf{w}_3) \text{ or } x + y + \sqrt{2}z = 0$$

$$0 = (\mathbf{u}_2, \mathbf{w}_3) \text{ or } x + y - \sqrt{2}z = 0.$$

One solution is

$$\mathbf{w}_3' = (1,\ -1,\ 0) \text{ or } \mathbf{u}_3' = \left(\frac{1}{\sqrt{2}},\ -\frac{1}{\sqrt{2}},\ 0\right).$$

Hence

$$A = \begin{bmatrix} \frac{1}{2} & \frac{1}{2} & \frac{1}{\sqrt{2}} \\ \frac{1}{2} & \frac{1}{2} & -\frac{1}{\sqrt{2}} \\ \frac{1}{\sqrt{2}} & -\frac{1}{\sqrt{2}} & 0 \end{bmatrix}.$$

This solution is not unique.

2.2 (1) Since $UU^* = I$

$$|a_1|^2 + |a_2|^2 = 1, \; |a_3|^2 + |a_4|^2 = 1$$

and

$$a_1 \bar{a}_3 + a_2 \bar{a}_4 = 0.$$

Similar conditions when $U^*U = I$ is considered.

(2) Let $\mathbf{u}_2' = (x, y, z)$ be the second column, then

$$x^2 + y^2 + z^2 + 1;$$

and

$$\frac{1}{2}x - \frac{1}{\sqrt{2}}iy + \frac{1}{2}z = 0,$$

then

$$\mathbf{u}_2' = \left(\frac{i}{2}, \frac{1}{\sqrt{2}}, \frac{i}{2}\right).$$

Let $\mathbf{u}_3' = (x, y, z)$ be the third column of U, then

$$\frac{1}{2}x + \frac{1}{\sqrt{2}}iy + \frac{1}{2}z = 0$$

and

$$-\frac{1}{2}ix + \frac{1}{\sqrt{2}}y - \frac{1}{2}iz = 0$$

Let $y = 0$, then $x = -z$ so

$$\mathbf{u}_3' = \left(\frac{1}{\sqrt{2}}, 0, -\frac{1}{\sqrt{2}}\right),$$

and

$$U = \begin{bmatrix} \frac{1}{2} & \frac{i}{2} & \frac{1}{\sqrt{2}} \\ \frac{i}{\sqrt{2}} & \frac{1}{\sqrt{2}} & 0 \\ \frac{1}{2} & \frac{i}{2} & -\frac{1}{\sqrt{2}} \end{bmatrix}$$

2.3 $\mathbf{z}_1 = \mathbf{y}_1 = (1, 1, 1)'$ so that

$$\mathbf{x}_1' = \left(\frac{1}{\sqrt{3}}, \frac{1}{\sqrt{3}}, \frac{1}{\sqrt{3}}\right).$$

$$\mathbf{z}_2' = \mathbf{y}_2' = (\mathbf{x}_1, \mathbf{y}_2)\mathbf{x}_1' = (1, 2, 3) - 2\sqrt{3}\left(\frac{1}{\sqrt{3}}, \frac{1}{\sqrt{3}}, \frac{1}{\sqrt{3}}\right).$$

$$= (-1, 0, 1)$$

Hence

$$\mathbf{x}_2' = \left(-\frac{1}{\sqrt{2}}, 0, \frac{1}{\sqrt{2}}\right).$$

$$\mathbf{z}_3' = \mathbf{y}_3' - (\mathbf{x}_2, \mathbf{y}_3)\mathbf{x}_2' - (\mathbf{x}_1, \mathbf{y}_3)\mathbf{x}_1'$$

$$= (0, 1, -1) + \frac{1}{\sqrt{2}}\left(-\frac{1}{\sqrt{2}}, 0, \frac{1}{\sqrt{2}}\right) - 0 = \left(-\frac{1}{2}, 1, -\frac{1}{2}\right)$$

so that

$$\mathbf{x}_3' = \left(-\frac{1}{\sqrt{6}}, \frac{\sqrt{2}}{\sqrt{3}}, -\frac{1}{\sqrt{6}}\right).$$

2.4 Let $A = PQ$, then

$$A^*A = Q^*P^*PQ = Q^*Q = I.$$

2.5 The matrix is nearly in the required form. We need only consider the submatrix

$$\begin{bmatrix} 2 & -\frac{3}{\sqrt{18}} \\ -\frac{6}{\sqrt{18}} & 2 \end{bmatrix}$$

which has an eigenvalue $\lambda = 1$ with the corresponding normalised eigenvector

$$\mathbf{x}' = \left(\frac{1}{\sqrt{3}}, \frac{\sqrt{2}}{\sqrt{3}}\right)$$

Using $\mathbf{x}$ as the first column, we add a second colum in such a way that the resulting matrix is unitary

$$W = \begin{bmatrix} \frac{1}{\sqrt{3}} & \frac{\sqrt{2}}{\sqrt{3}} \\ \frac{\sqrt{2}}{\sqrt{3}} & -\frac{1}{\sqrt{3}} \end{bmatrix}$$

so that

$$V = \begin{bmatrix} 1 & 0 & 0 \\ 0 & \frac{1}{\sqrt{3}} & \frac{\sqrt{2}}{\sqrt{3}} \\ 0 & \frac{\sqrt{2}}{\sqrt{3}} & -\frac{1}{\sqrt{3}} \end{bmatrix}$$

It is found that

$$V^*AV = \begin{bmatrix} -1 & \frac{10+\sqrt{2}}{6} & \frac{10\sqrt{2}-1}{6} \\ 0 & 1 & -\frac{1}{2} \\ 0 & 0 & 3 \end{bmatrix}.$$

2.6 (1) The eigenvalues of A are $\lambda = 0, 1, 4$. Corresponding to $\lambda = 1$, the (normalised) eigenvector is

$$\mathbf{x}' = (0, 0, 1).$$

Take

$$Q = \begin{bmatrix} 0 & 1 & 0 \\ 0 & 0 & 1 \\ 1 & 0 & 0 \end{bmatrix},$$

then

$$A^*AQ = \begin{bmatrix} 1 & 0 & 0 \\ 0 & 2 & 2 \\ 0 & 2 & 2 \end{bmatrix}.$$

The eigenvalues of

$$A_1 = \begin{bmatrix} 2 & 2 \\ 2 & 2 \end{bmatrix}$$

are $\mathbf{q} = 0$ and $\lambda = 4$.
Corresponding to $\lambda = 0$

$$\mathbf{x}' = \left(\frac{1}{\sqrt{2}}, -\frac{1}{\sqrt{2}}\right),$$

hence

$$V = \begin{bmatrix} 1 & 0 & 0 \\ 0 & \frac{1}{\sqrt{2}} & -\frac{1}{\sqrt{2}} \\ 0 & \frac{1}{\sqrt{2}} & \frac{1}{\sqrt{2}} \end{bmatrix}$$

so that

$$QV = \begin{bmatrix} 0 & \frac{1}{\sqrt{2}} & -\frac{1}{\sqrt{2}} \\ 0 & \frac{1}{\sqrt{2}} & \frac{1}{\sqrt{2}} \\ 1 & 0 & 0 \end{bmatrix}$$

and

$$(QV)^*A(QV) = \text{diag}\,\{1, 4, 0\}.$$

(2) Eigenvalues of A are $\lambda = 9, 9, -3$. Corresponding to $\lambda = 9$

$$\mathbf{x}' = \left(\frac{1}{\sqrt{2}}, \frac{1}{\sqrt{2}}, 0\right),$$

and

$$Q^*AQ = \begin{bmatrix} 9 & 0 & 0 \\ 0 & 1 & -\frac{8}{\sqrt{2}} \\ 0 & -\frac{8}{\sqrt{2}} & 5 \end{bmatrix}.$$

The eigenvalues of

$$A_1 = \begin{bmatrix} 1 & -\frac{8}{\sqrt{2}} \\ -\frac{8}{\sqrt{2}} & 5 \end{bmatrix}$$

are $\lambda = 9$ and $\lambda = -3$.

Corresponding to $\lambda = 9$,

$$\mathbf{x}' = \left(\frac{1}{\sqrt{3}}, -\frac{2}{\sqrt{3}}\right)$$

$$V = \begin{bmatrix} 1 & 0 & 0 \\ 0 & \frac{1}{\sqrt{3}} & \frac{\sqrt{2}}{\sqrt{3}} \\ 0 & -\frac{\sqrt{2}}{\sqrt{3}} & \frac{1}{\sqrt{3}} \end{bmatrix}$$

so that

$$QV = \begin{bmatrix} \frac{1}{\sqrt{2}} & \frac{1}{\sqrt{6}} & \frac{\sqrt{2}}{\sqrt{6}} \\ \frac{1}{\sqrt{2}} & -\frac{1}{\sqrt{6}} & -\frac{\sqrt{2}}{\sqrt{6}} \\ 0 & -\frac{\sqrt{2}}{\sqrt{3}} & \frac{1}{\sqrt{3}} \end{bmatrix}$$

and

$$(QV)^*A(QV) = \text{diag}\,\{9,\ 9,\ -3\}.$$

Note Considering $\lambda = -3$ first, we find

$$\mathbf{x}' = \left(\frac{1}{\sqrt{3}}, -\frac{1}{\sqrt{3}}, \frac{1}{\sqrt{3}}\right).$$

Then, using the Gram–Schmidt process on vectors

$$\mathbf{x}, \mathbf{e}_1 \text{ and } \mathbf{e}_2$$

we obtain

$$Q = \begin{bmatrix} \frac{1}{\sqrt{3}} & \frac{2}{\sqrt{6}} & 0 \\ -\frac{1}{\sqrt{3}} & \frac{1}{\sqrt{6}} & \frac{1}{\sqrt{2}} \\ \frac{1}{\sqrt{3}} & -\frac{1}{\sqrt{6}} & \frac{1}{\sqrt{2}} \end{bmatrix}$$

and

$$Q^*AQ = \text{diag}\,\{-3, 9, 9\}.$$

2.7 The right-hand side of the inequality has been proved in the text. We can write

$$\mathbf{x} = \sum \alpha_i \mathbf{x}_i$$

$$\phi(\mathbf{x}) = \frac{\lambda_1 |\alpha_1|^2 + \ldots + \lambda_n |\alpha_n|^2}{|\alpha_1|^2 + \ldots + |\alpha_n|^2}$$

so that

$$\phi(\mathbf{x}) - \lambda_n = \frac{(\lambda_1 - \lambda_n)|\alpha_1|^2 + \ldots + (\lambda_{n-1} - \lambda_n)|\alpha_{n-1}|^2}{|\alpha_1|^2 + \ldots + |\alpha_n|^2}$$

$$\geqslant 0$$

since

$$(\lambda_i - \lambda_n) \geqslant 0 \text{ all } i.$$

The left-hand side of the inequality follows.

2.8 (1) $$\phi(\mathbf{y}) = \frac{(\mathbf{x}_0 + \epsilon\mathbf{x})^* A(\mathbf{x}_0 + \epsilon\mathbf{x})}{(\mathbf{y}, \mathbf{y})}$$

Expanding and using the fact that

$$(\mathbf{x}_0, A\mathbf{x}) = (A\mathbf{x}_0, \mathbf{x})$$

and adding (and subtracting) the terms

$$\epsilon(\mathbf{x}_0, A\mathbf{x}) = \lambda_0(\mathbf{x}_0, \epsilon\mathbf{x})$$

and

$$(\epsilon\mathbf{x}, \epsilon\mathbf{x}) = |\epsilon|^2(\mathbf{x}, \mathbf{x}),$$

we obtain

$$\phi(\mathbf{y}) = \lambda_0 + \frac{|\epsilon|^2[(\mathbf{x}, A\mathbf{x}) - \lambda_0(\mathbf{x}, \mathbf{x})]}{(\mathbf{y}, \mathbf{y})}$$

Now using the identity

$$(\mathbf{x}, A\mathbf{x}) = (\mathbf{x}, \mathbf{x})\,\phi(\mathbf{x})$$

the result follows.

(2) $$\phi(\mathbf{y}) = \frac{4|y_1|^2 + 3i\bar{y}_1 y_2 - 3iy_1\bar{y}_2 + 2|y_2|^2}{|y_1|^2 + |y_2|^2}$$

where $y_1 = 1$ and $y_2 = -i$.

So

$$\phi(\mathbf{y}) = 6 \simeq \lambda_0.$$

Comment The eigenvalues of A are

$$\lambda_1 = 3 + \sqrt{10} = 6.16 \quad \text{and} \quad \lambda_2 = 3 - \sqrt{10} = -0.16.$$

Corresponding to λ_1, we have an eigenvector

$$\mathbf{x}_1' = (1, \ -0.72i).$$

The formula results in an apaproximation

$$\lambda_0 \simeq \phi(\mathbf{y})$$

which is accurate to a second order in ϵ, given $\mathbf{y}$ a first order approximation in ϵ to the eigenvector $\mathbf{x}_0$.

In general no approximation to an eigenvector is known. In such cases trial vectors can be used and good approximations can be found for the extreme eivenvalues λ_1 and λ_n. But this method is not useful for the intermediate eigenvalues $\lambda_2, \ldots, \lambda_{n-1}$.

2.9 Using Schur's inequality (Theorem 2.5)

$$\sum_{r=1}^{n} |\lambda_r|^2 \leqslant \sum_{i,j} |a_{ij}|^2$$
$$\leqslant n^2 \mu^2$$

(2) follows immediately.

Also by the above inequality

$$\frac{1}{n}\sum |\lambda_r|^2 \geqslant \pi |\lambda_r|^2$$
$$= |\det A|^{2/n} \qquad \text{(see 1.12)}$$

Hence

$$n\mu^2 \geqslant \frac{1}{n} \sum |\lambda_r|^2 \geqslant |\det A|^{2/n}$$

(1) follows.

2.10 We can write

$$C = \frac{C + C^*}{2} + \frac{C - C^*}{2}.$$

Let

$$H = \frac{C + C^*}{2},$$

then $H^* = H$, so H is Hermitian.

Let

$$S = \frac{C - C^*}{2},$$

then $S^* = -S$, so S is skew-Hermitian.

2.11 Assume $AB = BA$, then

$$(AB)^* = (BA)^* = A^*B^* = AB.$$

Conversely, assume that

$$(AB)^* = AB,$$

then

$$AB = (AB)^* = B^*A^* = BA.$$

2.12 (1) $\lambda_1 = 3,\ \lambda_2 = 1,\ \lambda_3 = -1.$
For

$$B_1 = B_2 = \begin{bmatrix} 0 & 0 \\ 0 & 3 \end{bmatrix},$$

the eigenvalues are

$$\mu_1 = 3 \quad \text{and} \quad \mu_2 = 0.$$

For

$$B_3 = \begin{bmatrix} 0 & -i \\ i & 0 \end{bmatrix},$$

$$\mu_1 = 1 \quad \text{and} \quad \mu_2 = -1.$$

So for each principal submatrix

$$\lambda_1 \geqslant \mu_1 \geqslant \lambda_2 \geqslant \mu_2 \geqslant \lambda_1.$$

(2) $\lambda_1 = 9,\ \lambda_2 = \lambda_3 = 3$

$$B_1 = B_2 = \begin{bmatrix} 5 & -2 \\ -2 & 5 \end{bmatrix}$$

has eigenvalues

$$\mu_1 = 7 \quad \text{and} \quad \mu_2 = 3.$$

B_3 also has eigenvalues $\mu_1 = 7$ and $\mu_2 = 3$.

2.13 (1) $[AA^*]^* = AA^*$. Similarly for A^*A.

(2) $\mathbf{xy}^* = \mathbf{yx}^*$ are matrices of order $(n \times m)$.

$$[\mathbf{xy}^* + \mathbf{yx}^*]^* = [\mathbf{yx}^* + \mathbf{xy}^*] = [\mathbf{xy}^* + \mathbf{yx}^*].$$

CHAPTER 3

3.1 (1) $$[xyz]\begin{bmatrix} 3 & -4 & 1/2 \\ -4 & -2 & 1 \\ 1/2 & 0 & 0 \end{bmatrix}\begin{bmatrix} x \\ y \\ z \end{bmatrix}$$

(2) $$[xyz]\begin{bmatrix} 1 & 1 & -2 \\ 1 & -2 & 3 \\ -2 & 3 & 3 \end{bmatrix}\begin{bmatrix} x \\ y \\ z \end{bmatrix}$$

3.2 (1) $\lambda = 1, 4, -2$

$$U = \begin{bmatrix} \frac{1}{\sqrt{3}} & \frac{2}{\sqrt{6}} & 0 \\ -\frac{1}{\sqrt{3}} & \frac{1}{\sqrt{6}} & -\frac{1}{\sqrt{2}} \\ -\frac{1}{\sqrt{3}} & \frac{1}{\sqrt{6}} & \frac{1}{\sqrt{2}} \end{bmatrix}$$

(2) $\lambda = 6, 6, 12$

$$U = \begin{bmatrix} \frac{1}{\sqrt{2}} & \frac{1}{\sqrt{3}} & \frac{1}{\sqrt{6}} \\ 0 & \frac{1}{\sqrt{3}} & -\frac{2}{\sqrt{6}} \\ -\frac{1}{\sqrt{2}} & \frac{1}{\sqrt{3}} & \frac{1}{\sqrt{6}} \end{bmatrix}$$

3.3 (1) Every eigenvalue of A is positive.

$$|A| = \lambda_1 \lambda_2 \ldots \lambda_n > 0,$$

hence A^{-1} exists.

Since

$$AA^{-1} = I$$

$$(A^{-1})'A' = (A^{-1})'A = I$$

So

$$(A^{-1})' = A^{-1};$$

hence A^{-1} is symmetric.

The eigenvalues of A^{-1} are

$$\lambda_1^{-1}, \lambda_2^{-2}, \ldots, \lambda_n^{-1},$$

and are all positive.

Hence A^{-1} is positive definite (see Theorem 3.2).

(2) The eigenvalues of A^m are

$$\lambda_1^m, \lambda_2^m, \ldots, \lambda_n^m,$$

they are all positive. The results follow from (1).

3.4 Consider $\mathbf{x} \neq \mathbf{0}$.

$$\mathbf{x}'[P+S]\mathbf{x} = \mathbf{x}'P\mathbf{x} + \mathbf{x}'S\mathbf{x}$$

Since

$$\mathbf{x}'P\mathbf{x} > 0 \quad \text{and} \quad \mathbf{x}'S\mathbf{x} \geqslant 0,$$

the result follows.

3.5 (1) $\Lambda = \text{diag}\{5, -3\}$, hence hyperbola.

(2) $\Lambda = \text{diag}\{2, -3\}$, hence hyperbola.

(3) $\Lambda = \text{diag}\{4, 6\}$, hence ellipse.

3.6 (1) Since $|3| > 0$,

$$\begin{vmatrix} 3 & -1 \\ -1 & 5 \end{vmatrix} = 14 > 0$$

and

$$\begin{vmatrix} 3 & -1 & 3 \\ -1 & 5 & 1 \\ 3 & 1 & 5 \end{vmatrix} = 16 > 0$$

A is positive definite.

(2) A is negative definite.

(3) A is positive semidefinite.

3.7 Since $A\mathbf{x} = \mathbf{y}$ (say) is a real vector

$$q(\mathbf{x}) = (A\mathbf{x})'A\mathbf{x} = \mathbf{y}'\mathbf{y}$$

where

$$\mathbf{y}' = [y_1 y_2, \ldots, y_n]$$

is a nonnegative scalar.

(1) $\mathbf{y}'\mathbf{y} = \sum_{i=1}^{1} y_i^2 > 0 \quad \text{all } \mathbf{x} \neq \mathbf{0},$

since all y_i $(i = 1, 2, \ldots, n)$ are linearly independent (remember the y_i are linear functions of the x_j).

(2) Again

$$\mathbf{y}'\mathbf{y} = \sum_{i=1}^{n} y_i^2$$

This time only r of the y_i $(i = 1, 2, \ldots, n)$ are linearly independent. So it is possible to select $\mathbf{x} \neq 0$ such that $\mathbf{y}'\mathbf{y} = 0$.

Hence $q(\mathbf{x})$ is semipositive definite.

3.8 Let $\mathbf{x} = U\mathbf{y}$.

$$q(\mathbf{x}) = \mathbf{x}'A\mathbf{x} = y'\Lambda y$$

But

$$\mathbf{y} = U'\mathbf{x} = \begin{bmatrix} (\mathbf{u}_1, \mathbf{x}) \\ (\mathbf{u}_2, \mathbf{x}) \\ \vdots \\ (\mathbf{x}_n, \mathbf{x}) \end{bmatrix} = \begin{bmatrix} \cos\theta_1 \\ \cos\theta_2 \\ \vdots \\ \cos\theta_n \end{bmatrix}$$

The result follows.

CHAPTER 4

4.1

	$p(x)$	$q(x)$
(1)	4	7
(2)	$\frac{11}{3}$	9
(3)	3	∞
(4)	$\frac{7}{2}$	∞
(5)	5	5

$$\max p(x) = 5 = \min q(x).$$

4.2 (1) $$|\lambda I - B| = \lambda^3 - 28\lambda^2 + 224\lambda - 512$$
$$= (\lambda - 16)(\lambda - 8)(\lambda - 4)$$

so that $\rho(B) = 16$.

$$|\lambda I - A| = \lambda^3 - 28\lambda^2 + 220\lambda - 496,$$

and

$$\rho(A) = 16.47.$$

(2) $\mathbf{x}$ corresponding to $\rho(B)$ is $[1 \ \ 1 \ \ 1]' > \mathbf{0}$.

4.3 (1) For matrix A

$$\sum_{j=1}^{n} a_{ij} = 16, \ 17, \ 16$$

Hence $16 \leqslant r \leqslant 17$.

(2) For matrix B

$$\sum_{j=1}^{n} a_{ij} = 16, \ 16, \ 16$$

hence $r = 16$.

4.4 $A\mathbf{x} = r\mathbf{x}$ and $A\mathbf{y} = \lambda\mathbf{y}$ but $\lambda \neq r$ (Theorem 4.7). Also $A'\mathbf{z} = r\mathbf{z}$ where $\mathbf{z} > \mathbf{0}$. By assumption that $\mathbf{y} \geqslant 0$, $(\mathbf{y}, \mathbf{z}) \neq 0$. Now

$$\lambda(\mathbf{y}, \mathbf{z}) = (\lambda\mathbf{y}, \mathbf{z}) = (A\mathbf{y}, \mathbf{z}) = (\mathbf{y}, A'\mathbf{z}) = (\mathbf{y}, r\mathbf{z}) = r(\mathbf{y}, \mathbf{z})$$

which is only possible if

$$(\mathbf{y}, \mathbf{z}) = 0.$$

The result follows.

4.5 (1) and (3) are reducible.
(2) and (4) are irreducible.

4.6 (1) Since $a_{j1}^{(3n+r)} = 1$ whenever $a_{j1}^{(r+3)} = 1$ the period is $d = 3$.

(2) $C_0 = \{1, 5\}$, $C_2 = \{2\}$ and $C_1 = \{3, 4, 6\}$.

(3) $$P = \begin{bmatrix} 0 & 0 & 1 & 0 & 0 & 0 \\ 0 & 0 & 0 & 1 & 0 & 0 \\ 0 & 0 & 0 & 0 & 0 & 1 \\ 0 & 1 & 0 & 0 & 0 & 0 \\ 1 & 0 & 0 & 0 & 0 & 0 \\ 0 & 0 & 0 & 0 & 1 & 0 \end{bmatrix}$$

(4) $$A_1 = [1\ 1\ 1],\ A_2 = \begin{bmatrix} 1 \\ 1 \end{bmatrix} \quad \text{and} \quad A_3 = \begin{bmatrix} 1 & 1 \\ 1 & 0 \\ 0 & 1 \end{bmatrix}.$$

(5) $$Q_3 = \begin{bmatrix} 2 & 2 & 2 \\ 1 & 1 & 1 \\ 1 & 1 & 1 \end{bmatrix},\ Q_2 = [4] \quad \text{and} \quad Q_1 = \begin{bmatrix} 2 & 2 \\ 2 & 2 \end{bmatrix}.$$

4.8 Since $\rho(Q_i) = 4$, $\rho(A) = 4^{1/3}$.

4.9 (1) $$P = \begin{bmatrix} 0 & 1 & 0 & 0 \\ 0 & 0 & 0 & 1 \\ 1 & 0 & 0 & 0 \\ 0 & 0 & 1 & 0 \end{bmatrix}.$$

(2) $$|\lambda I - A| = \lambda^4 - 3\lambda^3 - \lambda^2 + 5\lambda + 2$$

$$\rho(A) = \frac{2 + \sqrt{8}}{2} = 2.414.$$

(3) $$|\lambda I - A_{11}| = \lambda(\lambda^2 - 2\lambda - 1)$$

$$\rho(A_{11}) = 2.414$$

and

$$|\lambda I - A_{jj}| = (\lambda - 1)(\lambda^2 - \lambda - 2)$$

$$\rho(A_{jj}) = 2.414 \qquad (j = 2, 3, 4.).$$

CHAPTER 5

5.1 $$A^{-1} = \frac{1}{25}\begin{bmatrix} 10 & 10 & 5 \\ 5 & 15 & 5 \\ 5 & 10 & 10 \end{bmatrix} = \begin{bmatrix} 0.4 & 0.4 & 0.2 \\ 0.2 & 0.6 & 0.2 \\ 0.2 & 0.4 & 0.4 \end{bmatrix}.$$

5.2 $|\lambda I - A| = \lambda^3 - 6\lambda^2 + 10\lambda - 2 = (\lambda - 0.23)(\lambda^2 - 5.77\lambda + 8.673)$

so that

$$\lambda = 0.23,\ 2.885 \pm 0.591i.$$

5.3 $|B| = 46$

$$B^{-1} = \begin{bmatrix} 0.26 & 0.17 & 0.04 \\ 0.02 & 0.35 & 0.09 \\ 0.07 & 0.04 & 0.26 \end{bmatrix}.$$

5.4 $$L = \begin{bmatrix} 2 & 0 & 0 \\ 0 & 1 & 0 \\ -1/2 & -1/6 & 23/6 \end{bmatrix} \quad \text{and} \quad U = \begin{bmatrix} 2 & -1 & 0 \\ 0 & 3 & -1 \\ 0 & 0 & 1 \end{bmatrix}$$

(other solutions possible).

5.5 $$M^{-1} = \frac{1}{12}\begin{bmatrix} 3 & 1 & 0 \\ 0 & 4 & 0 \\ 0 & 1 & 3 \end{bmatrix} \geqslant 0 \quad \text{and} \quad N \geqslant 0$$

hence $M - N$ is a regular splitting.

$$C = M^{-1}N = \frac{1}{12}\begin{bmatrix} 0 & 3 & 1 \\ 0 & 0 & 4 \\ 3 & 3 & 1 \end{bmatrix}$$

$$|\lambda I - C| = \lambda^3 - \frac{1}{12}\lambda^2 - \frac{15}{144}\lambda - \frac{3}{144}.$$

$$\rho(C) \simeq 0.437.$$

5.6 $M_2^{-1} = \begin{bmatrix} 12 & 0 & 0 \\ 0 & 16 & 0 \\ 0 & 4 & 12 \end{bmatrix}$, $M_2^{-1}N_2 = E \text{ (say)} = \dfrac{1}{12}\begin{bmatrix} 0 & 8 & 0 \\ 0 & 0 & 4 \\ 3 & 3 & 1 \end{bmatrix}$

$$|\lambda I - E| = \lambda^3 - \frac{\lambda^2}{12} - \frac{\lambda}{12} - \frac{8}{144}$$

$$\rho(E) = 0.488$$

$$\rho(E) > \rho(C).$$

References

BOOK SECTION

[B1] Bellman, R. (1960) *Introduction to Matrix Analysis*, McGraw-Hill.

[B2] Berman, A. and Plemmons, R. J. (1979) *Nonnegative Matrices in the Mathematical Sciences*, Academic Press.

[B3] Cox, R. D. and Miller, H. D. (1965) *The Theory of Stochastic Processes* Methuen.

[B4] Davis, J. D. (1979) *Circulant Matrices*, John Wiley.

[B5] Franklin, J. N. (1968) *Matrix Theory*, Prentice-Hall.

[B6] Gantmacher, F. R. (1974) *The Theory of Matrices*, Vols. I and II, Chelsea.

[B7] Graham, A. (1979) *Matrix Theory and Applications for Engineers and Mathematicians*, Ellis Horwood.

[B8] Graham, A. and Burghes, D. (1980) *Introduction to Control Theory including Optimal Control*, Ellis Horwood.

[B9] Graham, A. (1981) *Kronecker Products and Matrix Calculus with Applications*, Ellis Horwood.

[B10] Hohn, F. E. (1973) *Elementary Matrix Algebra*, Macmillan.

[B11] Householder, A. S. (1953) *Principles of Numerical Analysis*, McGraw-Hill.

[B12] Householder, A. S. (1964) *The Theory of Matrices in Numerical Analysis*, Blaisdell.

[B13] Lancaster, P. (1969) *Theory of Matrices*, Academic Press.

[B14] Marcus, M. and Minc, M. (1964) *A Survey of Matrix Theory and Matrix Inequalities*, Allyn and Bacon.

[B15] Marcus, M. & Minc, M. (1965) *Introduction to Linear Algebra*, Macmillan.

[B16] Minc, M. (1978) *Encyclopedia of Mathematics and its Applications*, Addison-Wesley.

[B17] Noble, B. and Daniel, J. W. (1977) *Applied Linear Algebra*, Prentice-Hall.

[B18] Pullman, N. J. (1976) *Matrix Theory and Its Applications, Selected Topics*, Marcel Dekker.

[B19] Rorres, C. and Anton, H. (1984) *Applications of Linear Algebra*, John Wiley.

[B20] Schneider, H. (1964) *Recent Advances in Matrix Theory*, Madison and Wilwaukee, The University of Wisconsin Press.

[B21] Seneta, E. (1973) *Non-Negative Matrices. An Introduction to Theory and Applications*, George Allen & Unwin.

[B22] Turnbull, H. W. (1928) *Theory of Determinants Matrices and Invariants*, Blackie.

[B23] Varga, R. S. (1962) *Matrix Iterative Analysis*, Prentice-Hall Inc.

PAPERS SECTION

[1] Araki, M. (1975) Applications of M-Matrices to the Stability Problems of Composite Dynamical Systems, *J. of Math. Analysis and Appl.* **52**, 309–321.

[2] Barker, G. P., Berman, A. and Plemmans, R. J. (1978) Positive Diagonal Solutions to the Lyapunov Equations, *Linear and Multilinear Algebra,* **5**, 249–256.

[3] Brauer, A. (1957) A New Proof of Theorems of Perron and Frobenius on Non-Negative Matrices, *Duke Math. J.* **24**, 367–378.

[4] Brualdi, R. A., Parter, S. V. and Schneider, H. (1966) The Diagonal Equivalence of a Non negative Matrix to a Stochastic Matrix, *J. of Math. Analysis and Appl.* **16**, 31–50.

[5] Brualdi, R. A. and Hedricle, M. B. (1979) A Unified Treatment of Nearly Reducible and Nearly Decomposable Matrices, *Linear Algebra and Its Appl.* **24**, 51–73.

[6] Debreu, G. & Herstein, I. N. (1953) Non negative Square Matrices, *Econometrica* **21**, 597–607.

[7] Fiedler, M. & Ptak, V. (1962) On Matrices with Non-Positive Off-Diagonal Elements and Positive Principal Minors, *Czech. Math. J.* **12**, 382–400.

[8] Fiedler, M. and Ptack, V. (1966) Some Results on Matrices of Class K and their Application to the Convergence Rate of Iteration Procedures. *Czech. Math. J.* **16**, 260–272.

[9] Fiedler, M. & Ptak, V. (1967) Diagonally Dominant Matrices, *Czech. Math. J.* **17**, 420–433.

[10] Harary, F. (1959) Graph Theoretic Methods in the Management Sciences, *Management Science* **4**, 387–403.

[11] Hedrick, M. & Sinkhorn, R. (1970) A Special Case of Irreducible Matrices – The Nearly Reducible Matrices, *J. of Algebra* **16**, (143–150).

[12] Householder, A. S. (1958) On Matrices with Non-Negative Elements, *Montash. Math.* **62**, 238–242.

[13] Johnson, C. R. (1970) Positive Definite Matrices, *American Math. Monthly*, March, 259–264.

[14] Johnson, C. R. (1977) A Hadamard Product Involving M-Matrices, *Linear and Multilinear Algebra* **4**, 261–264.

[15] Johnson, C. R. and Robinson, H. A. (1981) Eigenvalue Inequalities for Principal Submatrices. *Linear Algebra and Its Appl.* **37**, 11–21.

[16] Johnson, C. R. (1982) Inverse M-Matrices, *Linear Algebra and Its Appl.* **47**, 195–216.

[17] Marcus, M. (1971) Linear Transformations on Matrices, *J. of Research of the Nat. Bur. of Standards* **75B**, 107–113.

[18] Minc, M. (1974) Linear Transformations on Nonnegative Matrices, *Linear Algebra and Its Appl.* **9**, 149–153.

[19] Minc, M. (1974) Irreducible Matrices, *Linear and Multilinear Algebra* **1**, 337–342.

[20] Ostrowski, A. M. (1960) On the Eigenvector Belonging to the Maximal Root of a Non-Negative Matrix, *Proc. Edinburgh Math. Soc.* **12**, 107–112.

[21] Plemmons, R. J. (1973) Regular Non negative Matrices, *Proc. Amer. Math. Soc.* **39**, 26–32.

[22] Plemmons, R. J. (1977) M-matrix Characterizations – Nonsingular M-Matrices, *Linear Algebra and Its Appl.* **18**, 175–188.

[23] Romanovsky, M. V. (1933) Un Théorème Sur Les Zéros des Matrices Non Négatives, *Bulletin. Soc. Math. de France* **61**, 213–219.

[24] Schneider, H. (1977) The Concepts of Irreducibility and Full Indecomposability of a Matrix in the Works of Frobenius, König and Markov, *Linear Algebra and Its Appl.* **18**, 139–162.

[25] Sinkhorn, R. & Knapp, P. (1967) Concerning Non negative Matrices and Doubly Stochastic Matrices, *Pacific J. of Math.* **21**.

[26] Sinkhorn, R. (1969) Concerning a Conjecture of Marshall Hall, *Proc. Amer. Math. Soc.* **21**, 197–201.

[27] Wielandt, H. (1950) Unzerlegbare, nicht negative Matrizen, *Math. Zeitschrift* **52**, 642–648.

Index

A CATALOG OF SELECTED

DOVER BOOKS

IN SCIENCE AND MATHEMATICS

Mathematics-Bestsellers

HANDBOOK OF MATHEMATICAL FUNCTIONS: with Formulas, Graphs, and Mathematical Tables, Edited by Milton Abramowitz and Irene A. Stegun. A classic resource for working with special functions, standard trig, and exponential logarithmic definitions and extensions, it features 29 sets of tables, some to as high as 20 places. 1046pp. 8 x 10 1/2. 0-486-61272-4

ABSTRACT AND CONCRETE CATEGORIES: The Joy of Cats, Jiri Adamek, Horst Herrlich, and George E. Strecker. This up-to-date introductory treatment employs category theory to explore the theory of structures. Its unique approach stresses concrete categories and presents a systematic view of factorization structures. Numerous examples. 1990 edition, updated 2004. 528pp. 6 1/8 x 9 1/4. 0-486-46934-4

MATHEMATICS: Its Content, Methods and Meaning, A. D. Aleksandrov, A. N. Kolmogorov, and M. A. Lavrent'ev. Major survey offers comprehensive, coherent discussions of analytic geometry, algebra, differential equations, calculus of variations, functions of a complex variable, prime numbers, linear and non-Euclidean geometry, topology, functional analysis, more. 1963 edition. 1120pp. 5 3/8 x 8 1/2. 0-486-40916-3

INTRODUCTION TO VECTORS AND TENSORS: Second Edition--Two Volumes Bound as One, Ray M. Bowen and C.-C. Wang. Convenient single-volume compilation of two texts offers both introduction and in-depth survey. Geared toward engineering and science students rather than mathematicians, it focuses on physics and engineering applications. 1976 edition. 560pp. 6 1/2 x 9 1/4. 0-486-46914-X

AN INTRODUCTION TO ORTHOGONAL POLYNOMIALS, Theodore S. Chihara. Concise introduction covers general elementary theory, including the representation theorem and distribution functions, continued fractions and chain sequences, the recurrence formula, special functions, and some specific systems. 1978 edition. 272pp. 5 3/8 x 8 1/2. 0-486-47929-3

ADVANCED MATHEMATICS FOR ENGINEERS AND SCIENTISTS, Paul DuChateau. This primary text and supplemental reference focuses on linear algebra, calculus, and ordinary differential equations. Additional topics include partial differential equations and approximation methods. Includes solved problems. 1992 edition. 400pp. 7 1/2 x 9 1/4. 0-486-47930-7

PARTIAL DIFFERENTIAL EQUATIONS FOR SCIENTISTS AND ENGINEERS, Stanley J. Farlow. Practical text shows how to formulate and solve partial differential equations. Coverage of diffusion-type problems, hyperbolic-type problems, elliptic-type problems, numerical and approximate methods. Solution guide available upon request. 1982 edition. 414pp. 6 1/8 x 9 1/4. 0-486-67620-X

VARIATIONAL PRINCIPLES AND FREE-BOUNDARY PROBLEMS, Avner Friedman. Advanced graduate-level text examines variational methods in partial differential equations and illustrates their applications to free-boundary problems. Features detailed statements of standard theory of elliptic and parabolic operators. 1982 edition. 720pp. 6 1/8 x 9 1/4. 0-486-47853-X

LINEAR ANALYSIS AND REPRESENTATION THEORY, Steven A. Gaal. Unified treatment covers topics from the theory of operators and operator algebras on Hilbert spaces; integration and representation theory for topological groups; and the theory of Lie algebras, Lie groups, and transform groups. 1973 edition. 704pp. 6 1/8 x 9 1/4. 0-486-47851-3

Mathematics–Algebra and Calculus

VECTOR CALCULUS, Peter Baxandall and Hans Liebeck. This introductory text offers a rigorous, comprehensive treatment. Classical theorems of vector calculus are amply illustrated with figures, worked examples, physical applications, and exercises with hints and answers. 1986 edition. 560pp. 5 3/8 x 8 1/2. 0-486-46620-5

ADVANCED CALCULUS: An Introduction to Classical Analysis, Louis Brand. A course in analysis that focuses on the functions of a real variable, this text introduces the basic concepts in their simplest setting and illustrates its teachings with numerous examples, theorems, and proofs. 1955 edition. 592pp. 5 3/8 x 8 1/2. 0-486-44548-8

ADVANCED CALCULUS, Avner Friedman. Intended for students who have already completed a one-year course in elementary calculus, this two-part treatment advances from functions of one variable to those of several variables. Solutions. 1971 edition. 432pp. 5 3/8 x 8 1/2. 0-486-45795-8

METHODS OF MATHEMATICS APPLIED TO CALCULUS, PROBABILITY, AND STATISTICS, Richard W. Hamming. This 4-part treatment begins with algebra and analytic geometry and proceeds to an exploration of the calculus of algebraic functions and transcendental functions and applications. 1985 edition. Includes 310 figures and 18 tables. 880pp. 6 1/2 x 9 1/4. 0-486-43945-3

BASIC ALGEBRA I: Second Edition, Nathan Jacobson. A classic text and standard reference for a generation, this volume covers all undergraduate algebra topics, including groups, rings, modules, Galois theory, polynomials, linear algebra, and associative algebra. 1985 edition. 528pp. 6 1/8 x 9 1/4. 0-486-47189-6

BASIC ALGEBRA II: Second Edition, Nathan Jacobson. This classic text and standard reference comprises all subjects of a first-year graduate-level course, including in-depth coverage of groups and polynomials and extensive use of categories and functors. 1989 edition. 704pp. 6 1/8 x 9 1/4. 0-486-47187-X

CALCULUS: An Intuitive and Physical Approach (Second Edition), Morris Kline. Application-oriented introduction relates the subject as closely as possible to science with explorations of the derivative; differentiation and integration of the powers of x; theorems on differentiation, antidifferentiation; the chain rule; trigonometric functions; more. Examples. 1967 edition. 960pp. 6 1/2 x 9 1/4. 0-486-40453-6

ABSTRACT ALGEBRA AND SOLUTION BY RADICALS, John E. Maxfield and Margaret W. Maxfield. Accessible advanced undergraduate-level text starts with groups, rings, fields, and polynomials and advances to Galois theory, radicals and roots of unity, and solution by radicals. Numerous examples, illustrations, exercises, appendixes. 1971 edition. 224pp. 6 1/8 x 9 1/4. 0-486-47723-1

AN INTRODUCTION TO THE THEORY OF LINEAR SPACES, Georgi E. Shilov. Translated by Richard A. Silverman. Introductory treatment offers a clear exposition of algebra, geometry, and analysis as parts of an integrated whole rather than separate subjects. Numerous examples illustrate many different fields, and problems include hints or answers. 1961 edition. 320pp. 5 3/8 x 8 1/2. 0-486-63070-6

LINEAR ALGEBRA, Georgi E. Shilov. Covers determinants, linear spaces, systems of linear equations, linear functions of a vector argument, coordinate transformations, the canonical form of the matrix of a linear operator, bilinear and quadratic forms, and more. 387pp. 5 3/8 x 8 1/2. 0-486-63518-X

Mathematics–Logic and Problem Solving

PERPLEXING PUZZLES AND TANTALIZING TEASERS, Martin Gardner. Ninety-three riddles, mazes, illusions, tricky questions, word and picture puzzles, and other challenges offer hours of entertainment for youngsters. Filled with rib-tickling drawings. Solutions. 224pp. 5 3/8 x 8 1/2. 0-486-25637-5

MY BEST MATHEMATICAL AND LOGIC PUZZLES, Martin Gardner. The noted expert selects 70 of his favorite "short" puzzles. Includes The Returning Explorer, The Mutilated Chessboard, Scrambled Box Tops, and dozens more. Complete solutions included. 96pp. 5 3/8 x 8 1/2. 0-486-28152-3

THE LADY OR THE TIGER?: and Other Logic Puzzles, Raymond M. Smullyan. Created by a renowned puzzle master, these whimsically themed challenges involve paradoxes about probability, time, and change; metapuzzles; and self-referentiality. Nineteen chapters advance in difficulty from relatively simple to highly complex. 1982 edition. 240pp. 5 3/8 x 8 1/2. 0-486-47027-X

SATAN, CANTOR AND INFINITY: Mind-Boggling Puzzles, Raymond M. Smullyan. A renowned mathematician tells stories of knights and knaves in an entertaining look at the logical precepts behind infinity, probability, time, and change. Requires a strong background in mathematics. Complete solutions. 288pp. 5 3/8 x 8 1/2. 0-486-47036-9

THE RED BOOK OF MATHEMATICAL PROBLEMS, Kenneth S. Williams and Kenneth Hardy. Handy compilation of 100 practice problems, hints and solutions indispensable for students preparing for the William Lowell Putnam and other mathematical competitions. Preface to the First Edition. Sources. 1988 edition. 192pp. 5 3/8 x 8 1/2. 0-486-69415-1

KING ARTHUR IN SEARCH OF HIS DOG AND OTHER CURIOUS PUZZLES, Raymond M. Smullyan. This fanciful, original collection for readers of all ages features arithmetic puzzles, logic problems related to crime detection, and logic and arithmetic puzzles involving King Arthur and his Dogs of the Round Table. 160pp. 5 3/8 x 8 1/2. 0-486-47435-6

UNDECIDABLE THEORIES: Studies in Logic and the Foundation of Mathematics, Alfred Tarski in collaboration with Andrzej Mostowski and Raphael M. Robinson. This well-known book by the famed logician consists of three treatises: "A General Method in Proofs of Undecidability," "Undecidability and Essential Undecidability in Mathematics," and "Undecidability of the Elementary Theory of Groups." 1953 edition. 112pp. 5 3/8 x 8 1/2. 0-486-47703-7

LOGIC FOR MATHEMATICIANS, J. Barkley Rosser. Examination of essential topics and theorems assumes no background in logic. "Undoubtedly a major addition to the literature of mathematical logic." – *Bulletin of the American Mathematical Society.* 1978 edition. 592pp. 6 1/8 x 9 1/4. 0-486-46898-4

INTRODUCTION TO PROOF IN ABSTRACT MATHEMATICS, Andrew Wohlgemuth. This undergraduate text teaches students what constitutes an acceptable proof, and it develops their ability to do proofs of routine problems as well as those requiring creative insights. 1990 edition. 384pp. 6 1/2 x 9 1/4. 0-486-47854-8

FIRST COURSE IN MATHEMATICAL LOGIC, Patrick Suppes and Shirley Hill. Rigorous introduction is simple enough in presentation and context for wide range of students. Symbolizing sentences; logical inference; truth and validity; truth tables; terms, predicates, universal quantifiers; universal specification and laws of identity; more. 288pp. 5 3/8 x 8 1/2. 0-486-42259-3

Mathematics–Probability and Statistics

BASIC PROBABILITY THEORY, Robert B. Ash. This text emphasizes the probabilistic way of thinking, rather than measure-theoretic concepts. Geared toward advanced undergraduates and graduate students, it features solutions to some of the problems. 1970 edition. 352pp. 5 3/8 x 8 1/2. 0-486-46628-0

PRINCIPLES OF STATISTICS, M. G. Bulmer. Concise description of classical statistics, from basic dice probabilities to modern regression analysis. Equal stress on theory and applications. Moderate difficulty; only basic calculus required. Includes problems with answers. 252pp. 5 5/8 x 8 1/4. 0-486-63760-3

OUTLINE OF BASIC STATISTICS: Dictionary and Formulas, John E. Freund and Frank J. Williams. Handy guide includes a 70-page outline of essential statistical formulas covering grouped and ungrouped data, finite populations, probability, and more, plus over 1,000 clear, concise definitions of statistical terms. 1966 edition. 208pp. 5 3/8 x 8 1/2. 0-486-47769-X

GOOD THINKING: The Foundations of Probability and Its Applications, Irving J. Good. This in-depth treatment of probability theory by a famous British statistician explores Keynesian principles and surveys such topics as Bayesian rationality, corroboration, hypothesis testing, and mathematical tools for induction and simplicity. 1983 edition. 352pp. 5 3/8 x 8 1/2. 0-486-47438-0

INTRODUCTION TO PROBABILITY THEORY WITH CONTEMPORARY APPLICATIONS, Lester L. Helms. Extensive discussions and clear examples, written in plain language, expose students to the rules and methods of probability. Exercises foster problem-solving skills, and all problems feature step-by-step solutions. 1997 edition. 368pp. 6 1/2 x 9 1/4. 0-486-47418-6

CHANCE, LUCK, AND STATISTICS, Horace C. Levinson. In simple, non-technical language, this volume explores the fundamentals governing chance and applies them to sports, government, and business. "Clear and lively ... remarkably accurate." – *Scientific Monthly*. 384pp. 5 3/8 x 8 1/2. 0-486-41997-5

FIFTY CHALLENGING PROBLEMS IN PROBABILITY WITH SOLUTIONS, Frederick Mosteller. Remarkable puzzlers, graded in difficulty, illustrate elementary and advanced aspects of probability. These problems were selected for originality, general interest, or because they demonstrate valuable techniques. Also includes detailed solutions. 88pp. 5 3/8 x 8 1/2. 0-486-65355-2

EXPERIMENTAL STATISTICS, Mary Gibbons Natrella. A handbook for those seeking engineering information and quantitative data for designing, developing, constructing, and testing equipment. Covers the planning of experiments, the analyzing of extreme-value data; and more. 1966 edition. Index. Includes 52 figures and 76 tables. 560pp. 8 3/8 x 11. 0-486-43937-2

STOCHASTIC MODELING: Analysis and Simulation, Barry L. Nelson. Coherent introduction to techniques also offers a guide to the mathematical, numerical, and simulation tools of systems analysis. Includes formulation of models, analysis, and interpretation of results. 1995 edition. 336pp. 6 1/8 x 9 1/4. 0-486-47770-3

INTRODUCTION TO BIOSTATISTICS: Second Edition, Robert R. Sokal and F. James Rohlf. Suitable for undergraduates with a minimal background in mathematics, this introduction ranges from descriptive statistics to fundamental distributions and the testing of hypotheses. Includes numerous worked-out problems and examples. 1987 edition. 384pp. 6 1/8 x 9 1/4. 0-486-46961-1